Springer Series on

# ATOMIC, OPTICAL, AND PLASMA PHYSICS

# PHYSICS                                         60

T0205708

Springer Series on

# ATOMIC, OPTICAL, AND PLASMA PHYSICS

The Springer Series on Atomic, Optical, and Plasma Physics covers in a comprehensive manner theory and experiment in the entire field of atoms and molecules and their interaction with electromagnetic radiation. Books in the series provide a rich source of new ideas and techniques with wide applications in fields such as chemistry, materials science, astrophysics, surface science, plasma technology, advanced optics, aeronomy, and engineering. Laser physics is a particular connecting theme that has provided much of the continuing impetus for new developments in the field. The purpose of the series is to cover the gap between standard undergraduate textbooks and the research literature with emphasis on the fundamental ideas, methods, techniques, and results in the field.

Please view available titles in *Springer Series on Atomic, Optical, and Plasma Physics* on series homepage *http://www.springer.com/series/411*

Afzal Chaudhry
Hans Kleinpoppen

# Analysis of Excitation and Ionization of Atoms and Molecules by Electron Impact

 Springer

Afzal Chaudhry
190 Barclay Court
Piscataway, New Jersey 08854
USA
chaudhryafzal@hotmail.com

Hans Kleinpoppen
Stirling University, UK
Max-Planck-Gesellschaft
Fritz-Haber-Institut
Faradayweg 4-6
14195 Berlin
Germany
kleinpop@fhi-berlin.mpg.de

Springer Series on Atomic, Optical, and Plasma Physics      ISSN 1615-5653
ISBN 978-1-4614-2713-1      ISBN 978-1-4419-6947-7 (eBook)
DOI 10.1007/978-1-4419-6947-7
Springer New York Heidelberg Dordrecht London

*Cover design:* eStudioCalamar S.L.

Printed on acid-free paper

Springer is part of Springer Science+Business Media (www.springer.com)

# Contents

# Chapter 1
# Introduction

**Abstract** An introduction to the study of electron impact excitation of hydrogen-like polarized atoms of sodium and potassium and helium and helium-like atoms of calcium and strontium is given. Also the introductory remarks for the investigation of the electron impact ionization of helium, argon, krypton and xenon atoms and hydrogen, sulphur dioxide and sulphur hexafluoride molecules is described.

**Keywords** Atoms · Excitation · Ionization · Molecules · Polarization

## 1.1 Ionization of Atomic Gases

The atomic collision process is, in general, a complex phenomenon and the interaction results in processes like elastic scattering, which dominates at low collision energies, excitation processes to bound and continuum states (ionization), electron exchange processes and correlation phenomenon of coherent impact as identified in the coincidence measurements. We are going to investigate first the ionization of atomic gases by electron impact.

The ionization of atoms by impact processes is of importance in a number of fields such as plasma physics, radiation physics, atmospheric physics, astrophysics and fusion physics (Loch et al. 2002) and in the study of the penetration of matter by electrons (see e.g. Bethe 1930, 1937; Massey and Mohr 1933; Fano 1963; Inokuti 1971). For many of these applications only total ionization data is needed. The experimental values of ionization cross sections, for atomic gases, of those by Nagy et al. (1980) and Schram (1966) are particularly noticeable. Kieffer and Dunn (1966) have made a compilation of the earlier experimental data from which it is clear that there is ~20% disagreement between the earlier experimental values of different research groups. Several basic methods have been used for calculations of ionization cross sections (Nagy et al. 1980). One method is based on the sum rule

A. Chaudhry and H. Kleinpoppen, *Analysis of Excitation and Ionization of Atoms and Molecules by Electron Impact*, Springer Series on Atomic, Optical, and Plasma Physics 60, DOI 10.1007/978-1-4419-6947-7_1, © Springer Science+Business Media, LLC 2011

which states that the difference between the total inelastic scattering cross sections $\sigma_{tot,inel}$, and the excitation cross section $\sigma_{exc}$, gives the ionization cross section $\sigma_{ion}$, (Inokuti et al. 1967; Inokuti 1971; Saxon 1973: Kim et al. 1973; Eggarter 1975). Other theoretical treatments are based on the dispersion-relation analysis of electron-atom scattering (Bransden and McDowell 1969, 1970; de Heer et al. 1979). For this analysis, accurate values of the total scattering cross section, $\sigma_{tot}$, are very important (de Heer and Jansen 1977; de Heer et al. 1979), since use is made of the relation $\sigma_{tot} = \sigma_{exc} + \sigma_{ion} + \sigma_{el}$, where $\sigma_{el}$ is the total elastic scattering cross section.

For a number of applications the knowledge of the energy and the angular distribution of the electrons produced by the collision process (variously referred to as ejected or secondary electrons) are also necessary. In this case, from a theoretical point of view the basic quantity which is calculated is a triple differential cross section (TDCS), differential in the solid angle of the emitted electrons, characterized by their energy. The integration of the TDCS over the electron energy yields the double differential cross section (DDCS), that is the ionization cross section differential in energy and the angular distribution of the electrons produced in the ionization.

We like to refer and restrict ourselves to a number of the measurements of the DDCS which have been published (Ehrhardt et al. 1972; Peterson et al. 1971, 1972; Opal et al. 1971, 1972; Vroom et al. 1977; Oda et al. 1972; Tahira and Oda 1973; Shyn et al. 1979, 1981; Kim Y.-K. (1983). Theoretical studies of DDCS in the First Born Approximation (FBA) (Madison 1973; Bell and Kingston 1975); Plane Wave Born Approximation (PWBA) (Tahira and Oda 1973, and references therein) and Binary Encounter Theory (BET) (Oda et al. 1972; Vriens 1969; Bonsen and Vriens 1970) have been reported.

Another aspect of the atomic collision process, which is of considerable interest, is the multiple ionization of the atom, which can be produced by several processes such as the following:

(a) Direct multiple ionization
(b) Multiple-ionization involving correlation between electrons
(c) Ionization of the inner shells followed by Coster-Kronig and/or Auger transitions
(d) Ionization followed by a core relaxation process (shake-off)

The investigations of the multiple ionization process have employed direct detection of the multiply charged ions (Van der wiel and Wiebes 1971; Shram 1966; Nagy et al. 1980) or have used indirect means such as the detection of vacuum ultraviolet radiation (Beyer et al. 1979), Auger electrons (Stolterfoht et al. 1973) or Characteristic X-rays (Oona 1974). In these studies total cross sections for the production of multiple-ionization are measured.

In the present work, however, we have investigated the atomic collision process (with emphasis on electron-atom collisions) in more detail by measuring double differential cross sections for $n$-fold ionization or partial double differential cross sections DDCS($n+$), i.e. ionization cross sections for multiple ionization differential in secondary electron energy and its ejection angle, using the electron-ion coincidence technique.

The experimental set-up for measuring DDCS ($n+$) by the electron-ion coincidence technique is given in Sect. 4.1. In these investigations electrons ejected at $90°$ to the incident electron direction are energy analyzed and then detected in coincidence with the product ions which are analyzed by a time-of-flight (TOF) type analyzer. The apparatus used in these experiments is discussed in Sects. 3.1.1–3.1.6. Relative values of the DDCS ($n+$) and the DDCS (obtained by summing the DDCS ($n+$) over all values of $n$) have been measured for helium, argon, krypton and xenon as a function of the secondary electron energy and the incident electron energy. The results have been discussed in Sects. 5.1–5.1.4. Present DDCS values have been compared with other similar experimental data from literature. Comparisons, with the theoretical predictions, have also been given, where possible. In the absence of a comprehensive theoretical explanation for multiple ionization (McGuire 1982), the values of DDCS ($n+$) for these many electron atoms cannot be checked by theory. For neon a single measurement is reported in Sect. 5.1.4.

The experimental set-up for the X-ray-ion coincidence experiment, investigating the electron-xenon-atom collision process, is given in Sect. 4.1.6. In this experiment X-rays are detected by a liquid-nitrogen-cooled hyperfine germanium (HPGe) detector while ions are analyzed for charge state by a time-of-flight (TOF) type analyzer. The apparatus used for this experiment is described in Sect. 3.1.9. The results are discussed in Sect. 5.1.5.

A plane crystal X-ray spectrometer (Harbach 1980; Werner 1983; Jitschin 1984) has been used to study $K_\alpha$ and $K_\beta$ X-ray lines emitted by $^{54}Mn$ as a result of electron capture (EC) in a $^{55}Fe$ radioactive source. The apparatus used in this experiment is described in Sect. 3.1.12 and the experimental set-up, including the circuit for interfacing with a micro-computer, is given in Sect. 4.3. The results of this experiment are discussed in Sect. 5.1.6.

The areas of research described in this book outline the developments, which are mainly, selectively, covered by data and techniques for educational and informative instructions. A more comprehensive description of present achievements and summaries can be found in the work and book on "Cold Target Recoil Ion Momentum Spectroscopy and Reaction Microscopes "edited by J. Ullrich in 2004 (J. Ullrich, Max-Planck Gesellschaft, Heidelberg, April 2004). It includes Bench Mark Experiments with heavy projectiles, single photons, electrons, weak lasers and pump-probe experiments, and intense and ultra-short laser pulses. Reviews include topics on "Cold Target Recoil-Ion Momentum Spectroscopy," "Multiple Ionization in Strong-Laser-Field" and "Reaction Microscopes," which are published by the Max Planck Society. It also completely lists articles on all possible aspects of the physics of atomic and molecular ionization processes from 1994 to 2004 including kinematically complete type of collision experiments (excluding quantum mechanically complete collision experiments). Although the books by U. Becker and D. A. Shirley (1996) and Beyer et al. (1997) are somewhat out of date yet they are being referred to here as well.

The theoretical treatment for these investigations is outlined in Sects. 2.1–2.1.9, the apparatus is described in Sects. 3.1.1–3.1.12.6, the experimental techniques used are given in Sects. 4.1–4.3, the measurements are described in the Sects. 5.1–5.1.6 and the concluding remarks are given in Chap. 6.

## 1.2 Ionization of Hydrogen, Sulphur Dioxide and Sulphur Hexa Fluoride

The present measurements for electron impact dissociative and non-dissociative ionization of molecular gases such as $H_2, SO_2$ and $SF_6$ are described in Sects. 5.2.1–5.4.3. The importance and the consequences of these processes have been recognized in the modelling of planetary and cometary atmospheres (Nier 1985; Zipf 1985), and in the various devices such as glow discharge lamps, lasers and gaseous switches (Mark and Dunn 1985). The hydrogen molecule being the simplest molecule has been extensively studied both experimentally and theoretically and investigations have been made for single- (Kossmann et al. 1990; Edwards et al. 1988; Rapp et al. 1965; Tawara et al. 1990), double- (Kossmann et al. 1990; Edwards et al. 1988; Tawara et al. 1990; Edwards et al. 1989; McCulloh et al. 1968), and dissociative single- (Kossmann et al. 1990; Tawara et al. 1990; Dunn et al. 1963; Cho et al. 1986; Crowe et al. 1973; Kieffer et al. 1967; Van Brunt et al. 1970; Brehm et al. 1978; Kollmann 1975, 1978; Crowe et al. 1973b) ionization cross sections of molecular hydrogen by electron impact. Also, measurements of singly and doubly differential cross sections for electron ejection, which do not discriminate between single or double ionization, have been reported in the literature (Opal et al. 1972; DuBois and Rudd 1978; Shyn et al. 1981). Section 5.2.1 describes the present measurements of partial double differential cross sections (PDDCS) for molecular single ionization [PDDCS($H_2^+$)], dissociative single ionization [PDDCS(H+$H^+$)] or more simply [PDDCS($H^+$)] and doubly differential cross sections (DDCS) [$= $ PDDCS($H_2^+$) plus PDDCS($H^+$)].

Sulphur dioxide ($SO_2$) is one of the most abundant pollutants being released into the atmosphere, especially in cities and around large industrial and power plants where it results, from the combustion of fossil fuels (Barker 1979; Cadez et al. 1983). It is well known that sulphur dioxide is largely responsible for the phenomenon of acid rain which is causing large areas of forests and lakes to lose their ability to support animal and plant life (Cooper et al. 1991a; Bridgeman 1991). The interest in the electron impact dissociation processes in $SO_2$ has been stimulated recently by the suggested use of electrical discharges as a means of destruction and removal of this molecule from the exhaust gases of electricity generation stations (Burgt et al. 1992; Penetrante et al. 1991). The experimental electron impact ionization data of $SO_2$ molecule are of great interest in the modelling of Jupiter-Io's atmosphere (Cheng 1980) and of the plasma of diffuse-discharge switches (see Orient and Srivastava 1984). There is, therefore, considerable environmental, astronomical as well as fundamental scientific interest in the processes responsible for the dissociation of $SO_2$ molecule by electron impact (Reese et al. 1958; Smith and Stevenson 1981).

Section 5.2.2 describes the electron impact ionization of $SO_2$ molecule and gives the present measurements of doubly differential cross sections (DDCS) for the ionization of $SO_2$ and partial double differential cross sections (PDDCS) for the ions resulting from the dissociation of $SO_2$ molecule by electron impact for different incident electron energies. Measurements are also given for the angular variation of DDCS for the ionization of $SO_2$ molecule and PDDCS for the ions resulting from the dissociation of the $SO_2$ molecule at different incident electron energies. These data have also been transformed into percentage branching ratios (BR%) to enable a comparison with similar data published in literature (Cooper et al. 1991a; Cooper et al. 1991b; Orient and Srivastava 1984; Smith and Stevenson 1981).

The ionization properties of the $SF_6$ molecule are particularly interesting since it was used, for example, for the first separation of sulphur isotopes by laser irradiation in 1975 (Ambartsumyam et al. 1975) and as a model for the study of laser-induced chemistry, and acts as an electron scavenger in gaseous discharges due to the large cross section for the formation of $SF_6$ ions at near-zero electron energies (Fehsenfeld 1970; Lifshitz et al. 1973; Compton et al. 1978). Owing to the properties of the $SF_6$ molecule, namely its high dielectric strength, its chemical inertness and its high saturation vapour pressure at room temperature, it has received a wide acceptance in industry as a gaseous insulator in high-voltage electrostatic generators, transformers, condensers and cables (Hauschild and Exner 1987; Johnstone and Newell 1991; Talib and Saporoschenko 1992) and is also used in a plasma-etching technique which is important in reducing large scale integrated circuits to sub-micron levels (Endo and Kurogi 1980; Flamm and Donnelly 1981; Coburn 1982; Pinto et al. 1987; Roque et al. 1991). A thorough understanding of the nature and properties of the decomposition products of $SF_6$ under electron bombardment is, therefore, a matter for primary concern. The $SF_6$ molecule has, in fact, been studied by many workers and the partial dissociative ionization cross sections for $SF_x^{n+}$ ions with $x = 0$–$5$, $n = 0$–$4$ and also for the $F^+$ ion have been measured (Dibeler and Mohler 1948; Marriott 1954; Dibeler and Walker 1966; Harland and Thynne 1969; Pullen and Stockdale 1976; Hitchcock et al. 1978, Hitchcock and Van der Wiel 1979; Masuoka and Samson 1981, Stanski and Adamczyk 1983, Margreiter et al. 1990). The coincidence method used by these authors involved an electron–ion (e, e+ ion) coincidence technique but their measurements were only for non-differential cross sections. However, no electron–electron (e, 2e) coincidence measurements, which would lead to the measurements of doubly (DDCS) or triply (TDDCS) differential cross sections, have been carried out on the $SF_6$ molecule. This fact may be due to the high resolution of the equipment required to obtain the (e, 2e) coincidences for molecules like $SF_6$.

Section 5.2.3 describes the electron impact ionization of $SF_6$ molecule and gives the measurements of doubly differential cross sections (DDCS) for the ionization of $SF_6$ and partial double differential cross sections (PDDCS) for ions such as $SF_5^+$, $SF_4^+$, $SF_3^+$, $SF_2^+$, $(SF^+, SF_4^{++})$, $S^+$, $SF_5^{++}$, $SF_3^{++}$ and $SF_2^{++}$ resulting from the dissociative ionization of $SF_6$ molecule by electron impact.

The theoretical treatment of the electron impact ionization of molecular gases is given in Sect. 2.2, the apparatus used in these experiments is described in Sect. 3.2, the experimental arrangement is shown schematically in the Sect. 4.4, the measurements of the double differential cross sections for the dissociative and non-dissociative ionization of molecular gases by electron impact are given in Sects. 5.2.1–5.2.3 and the concluding remarks are mentioned in the Chap. 6.

## 1.3   Excitation of Polarized Sodium and Potassium Atoms

Sodium and potassium atoms are composed of a single electron outside a core of completely filled electron shells and the outer electron is in the nS state with $n = 3$ for sodium and $n = 4$ for potassium. These atoms are considered to be in excited state if the outer electron is found to be in the 3P or any higher state for sodium and in the 4P or any higher state for potassium. The electron impact nS-nP excitation of alkali metal atoms has been the subject of numerous experimental and theoretical investigations [refer to the reviews by Moisewitch and Smith (1968, 1969) and Bransden and McDowell (1977, 1978)]. The total cross sections (Moisewitch and Smith (1968)) for nS-nP excitations of sodium and potassium atoms, due to the strong coupling between the initial and final states of the resonance transitions, have the highest values in the neighbourhood of the excitation thresholds.

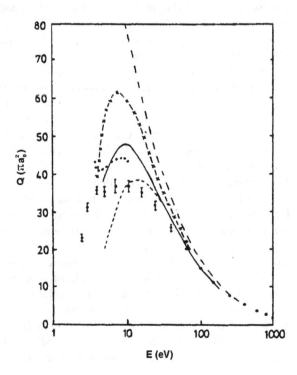

**Fig. 1.1**  Integrated (total) cross section ($\pi\, a_o^2$) for the first resonance transition of NaI up to 1 keV incident electron energy. Experimental results; (•) Enemark and Gallagher (1972); (x-x-) Zapesochnyi et al. (1975). Theoretical results: (-.-.-.-) FBA (Walters 1973); (- - - -) Glauber (Walters 1973); (.......) McCavert and Rudge (1972); (- - - - - - - -) UNWPO II (Kennedy et al. 1977)

**Fig. 1.2** Integrated (total) cross section ($\pi\, a_o^2$) for the first resonance transition of KI up to 1 keV. Experimental results; (o) William and Trajmar (1977); (x-x-) Zapesochnyi et al. (1975). Theoretical results: (-.-.-.-) FBA (Walters 1973); (- - - -) Glauber (Walters 1973); (. . . . . .) McCavert and Rudge (1972); (- - - - - - - -) UNWPO II (Kennedy et al. 1977)

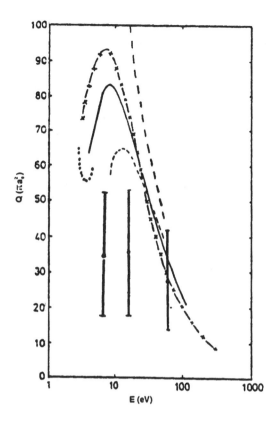

Figures 1.1 and 1.2 show (Heddle and Gallagher 1989) a comparison of experimental and theoretical works to find the optical excitation functions of the resonance radiation of sodium and potassium atoms. To find the optical excitation functions of the resonance radiation of the sodium and potassium atoms, the measurement of the total excitation cross is needed because when atoms are excited by the un-polarized electrons and the light emitted in the subsequent decay is observed without detecting the scattered electrons, then the polarization properties of the emitted radiation depend on two parameters, the total cross section and the atomic alignment (Bartchat and Blum 1982; Percival and Seaton 1958). In such a measurement if we choose the incoming beam shell to be in the - OY direction then the radiation emitted may be considered to be due to the electric dipoles in the OY direction and the two other equal dipoles in the OX and OZ directions. Using unpolarized electrons and polarized atoms Enemark and Gallagher (1972) measured the total cross sections for nS-nP excitation and found that at low energy, from threshold up to 5 eV, the normalized cross section and the polarization are in excellent agreement with the close coupling calculations. Kennedy et al. (1977) have compared their calculations for the total cross sections of the excitation of the first resonance lines of sodium and potassium atoms using the

unitarized distorted-wave polarized-orbital (UNWPO) model with other experimental and theoretical works from literature. Heddle and Gallagher (1989) have compared, in their review paper, the experimental work for the measurement of the optical excitation functions for atoms of sodium and potassium (see Figs. 1.3 and 1.4).

So far, we have given the total cross sections and the optical excitation functions when both colliding partners are unpolarized. For a better understanding of the behaviour of these functions, knowledge of the different channels of interaction is essential. Table 1.1 gives the different interaction channels involved in the scattering process of the unpolarized partners.

One of the collision partners has to be polarized to enable the calculation of the contributions of different interaction channels to the total cross section. We have used polarized atoms of sodium and potassium as targets and, therefore, the different interaction channels are as given below:

$$e(\uparrow\downarrow) + A(\uparrow) \rightarrow e(\downarrow) + A^*(\uparrow) \text{ direct,} \tag{1.1}$$

$$e(\uparrow\downarrow) + A(\uparrow) \rightarrow e(\uparrow) + A^*(\downarrow) \text{ exchange,} \tag{1.2}$$

$$e(\uparrow\downarrow) + A(\uparrow) \rightarrow e(\uparrow) + A^*(\uparrow) \text{ interference,} \tag{1.3}$$

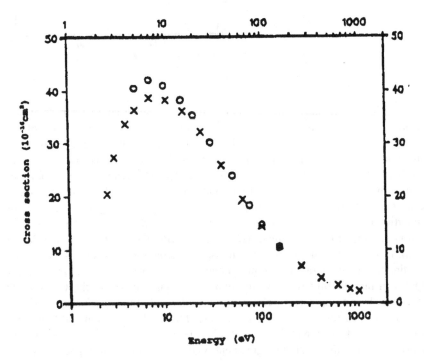

**Fig. 1.3** Excitation function of the D-lines of sodium: (x) Enemark and Gallagher (1972); (o) Phelps and Lin (1981)

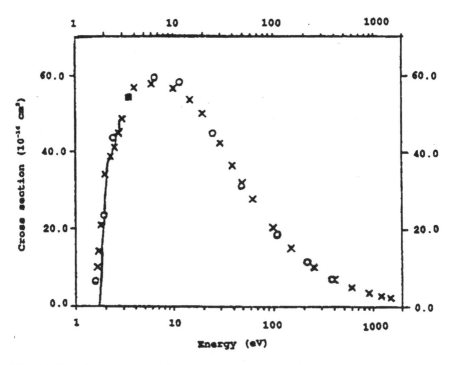

**Fig. 1.4** Excitation function of the D-lines of potassium: (x) Chen and Gallagher (1978); (o) Phelps et al. (1979); (- - - - -) Papp et al. (1983)

**Table 1.1** Electron scattering by spin ½ atoms

| Component of the electron spin is along the magnetic field | | | | |
|---|---|---|---|---|
| Before collision | | After collision | | Channel of interaction |
| Atomic electron | Incident electron | Atomic electron | Scattered electron | |
| ½ | ½ | ½ | ½ | Interference |
| ½ | −½ | ½ | −½ | Direct |
| ½ | −½ | −½ | ½ | Exchange |
| −½ | ½ | ½ | −½ | Exchange |
| −½ | ½ | −½ | ½ | Direct |
| −½ | −½ | −½ | −½ | Interference |

where the arrows indicate the spin direction, $e(\uparrow\downarrow)$ stands for unpolarized electron, $A(\uparrow)$ for polarized atom and $A^*$ for the excited atom.

In this experiment, the photons are detected along the OZ-axis with the atomic beam in the OX direction. If the atomic polarization vector is along the OZ direction then the experiment does not have a cylindrical symmetry about the electron beam direction and the circular polarization of the observed decay light is not zero

(Bartschat and Blum 1982). An investigation of the properties of light from the collision of the polarized partners was first proposed by Kleinpoppen (1971). Although for impact energies close to the threshold all reaction channels have comparable magnitudes but at higher energies the exchange interaction becomes negligible. If the spin of the atomic collision partner is known prior to and after the collision then the information about the exchange interaction can be obtained directly. The exchange cross section for the scattering of low energy electrons by potassium has been measured by Ruben et al. (1960). Campbell et al. (1971) used exchange interaction to polarize electrons by scattering them from the polarized potassium atoms. Schröder (1982) and Baum et al. (1983) investigated the interference between the direct and exchange interactions by using polarized electrons and polarized sodium atoms. Jitschin et al. (1984) were the first to report a polarization analysis of the fluorescence light from the excitation of spin polarized sodium atoms by electron impact.

A brief theoretical discussion of the fluorescence analysis is given in Sect. 2.3, the apparatus used is described in Sects. 3.3.1–3.3.8, the experimental technique is outlined in Sects. 4.5–4.10, the results and discussion are included in Sects. 5.3.1–5.3.2 and the concluding remarks are given in Chap. 6.

## 1.4   Excitation of Alkaline Earth Metal Atoms of Calcium and Strontium

Since the pioneering electron-photon correlation studies (Eminyan et al. 1973, 1974; Standage and Kleinpoppen 1976) of the electron impact excitation of $^1$P states of helium the two-electron system of helium has played a dominant role in such complete experiments, and a wealth of experimental and theoretical knowledge has now been accumulated for the excitation of the various helium states (for a review, see Andersen et al. 1988). It is of particular interest, however, to study heavier atoms with a helium-like two-electron outer shell such as alkaline earth metal atoms of Be, Mg, Ca, Sr and Ba. The ground and excited states of the lower members of this series are, like He, well described by $LS$ coupling. However, the fact that there are closed electron shells below the outer $ns^2$ shell which can be excited by the electron impact, leads to the modifications of the state structure and of the electron scattering process. Compared with helium the lowest excited $^1$P states of alkaline earth metals are much more strongly coupled to the ground state and electron correlation effects are, therefore, expected to play a major role in the electron-impact excitation process (Robb 1974; Sadlej et al. 1991). The influence of these atomic configurations on the scattering parameters and a test of the dynamical collision approximations are the main motivation for the electron impact collision studies of these atoms.

## 1.4.1 Excitation-Ionization of the Calcium Atom by Electron Impact

The electron impact excitation of ionic states from ground-state atoms at low target pressure are usually a single-step process of simultaneous ionization and excitation. The corresponding cross sections are small and a sophisticated theoretical treatment is required to describe these processes. Nevertheless, the cross sections and the excitation functions have been studied experimentally and theoretically for a number of atoms, most extensively for helium where excitation cross sections into various He II states have been calculated by Dalgarno and McDowell (1956), Lee and Lin (1965), Gillespie (1972). Kheifets et al. (1999) and Fang and Bartschat (2000), and measurements have been reported by Elenbaas (1930), Hughes and Weaver (1963), Haidt and Kleinpoppen (1966), Anderson et al (1967), Anderson and Hughes (1972), Forand et al. (1985) and Avaldi et al. (1998). Because of the l degeneracy of hydrogen-like atoms it is difficult to extract the contributions of individual states from the measurements. Forand et al. (1985) have avoided this problem in the study of the He II $\lambda = 30.4$ nm (2P→1S) transition. Fang and Bartschat (2000) have reported convergent second-order calculations for simultaneous electron-impact ionization-excitation of helium. Little work seems to have been done for the investigation of simultaneous ionization and excitation of other atoms except for the measurements of Mg II (Leep and Gallagher 1976), Ba II (Chen and Gallagher 1976; Goto et al. 1983) and Zn II (Inaba et al. 1986).

More information about the simultaneous ionization and excitation is provided by the polarization of the emitted spectral line and also it is of interest to see if the theory of Percival and Seaton (1958) which predicts the polarization of atomic transitions following electron impact excitation does apply to this more complicated process. The polarization measurements have been carried out by Elanbaas (1930) and Haidt and Kleinpoppen (1966) for the He II $\lambda = 468.6$ nm ($n = 4 \rightarrow n = 3$) line complex, Leep and Gallagher (1976), Chen et al. (1976) and Chen and Gallagher (1976) for the lowest P $\rightarrow$ S transitions for Mg II, Sr II and Ba II and Goto et al. (1983) for the low-lying states of Cd II.

Stevenson and Crowe (2004) have studied the excitation-ionization of the calcium atom by electron impact and have reported preliminary measurements of linear Stokes parameters for the 4p $^2P_{3/2}$ excited state of Ca$^+$ produced from 4s$^2$ $^1$S ground state of the calcium atom. Shintarou Kawazoe et al. (2006) have reported R-matrix calculations for the excitation of the 4 $^1P^0$ state of Ca from the ground 4 $^1$S state and conclude that their calculations agree with the experimental and other theoretical calculations, especially at lower energies.

The excitation function and the polarization of the Ca II $\lambda = 393.3$ nm (4 $^2P_{3/2} \rightarrow 4$ $^2S_{1/2}$) line following electron impact simultaneous ionization and excitation from the Ca I ground state have been measured in the present study from threshold to 60 eV for the excitation function and to 200 eV for the polarization measurements. The absolute excitation cross section of the Ca II 4 $^2P_{3/2}$ state has also been measured at the incident electron energy of 40 eV.

Sections 3.4, 4.11.1, 5.4.1 and Chap. 6 give the apparatus, experimental technique, results and discussion and the concluding remarks, respectively.

### 1.4.2   Measurement of Coherence and Polarization Parameters for the Excitation of the 5 $^1$P State of Strontium

Brunger et al. (1987, 1989) measured electron-photon angular and polarization correlation for the electron impact excitation of 3 $^1$P state of Mg and corresponding calculations have been reported by Mitroy and McCarthy (1989), Meneses et al. (1990) and Clark et al. (1991). Clark et al. (1989) have also reported their calculations for the electron impact coherence parameters for the 6 $^1$P state of Barium.

The electron impact excitation of the 5 $^1$P state in strontium has presently been studied by the electron-photon coincidence technique. The Stokes parameters $P_1$, $P_2$ and $P_3$ of the decay transition (5 $^1P_1 \rightarrow$ 5 $^1S_0$) at $\lambda = 460.7$ nm have been measured for an incident electron energy of 45 eV. The present measurements extend over the electron scattering angles 30–113°.

Sections 3.5, 4.11.2, 5.4.4 and Chap. 6 give the apparatus, experimental technique, results and discussion and the concluding remarks, respectively, for the present strontium measurements.

## 1.5   Polarization Correlation Measurements of 3 $^1$P State of Helium

Inelastic electron-atom scattering processes usually transfer alignment and orientation to the atoms. If alignment and orientation are measured, for example in coincidence with the scattered electron, the scattering amplitudes can be determined completely (apart from an overall phase factor). To determine the sign of the orientation a measurement of the circular polarization of the photon emitted by the excited atom is required, but the alignment and the absolute value of the orientation can be derived either from the linear polarization measurements or from the angular correlation measurements made on the system to extract all scattering parameters.

Standage and Kleinpoppen (1976) were the first to measure the complete scattering parameters including the sign of $<L>$ on the 3 $^1$P state of helium using 80 eV incident electrons and scattering angles between 15° and 27.5°. The interest in the sign of $<L>$ in connection with the scattering models (Steph and Golden 1980; Kohmoto and Fano 1981; Madison et al. 1986) has encouraged a series of polarization measurements on the 2$^1$P (Williams 1983; Khakoo et al. 1986) and 3 $^1$P states of helium (Ibrahiem et al. 1985; Beijers et al. 1986). Crowe and Rudge (1988) have also reported measurements on the 2,3 $^1$P states of helium. Fano et al. (1991) have published their R-matrix calculations for the $n$ $^{3,1}$P ($n = 2$–4) states of helium.

We are reporting polarization correlation measurements for 3 $^1$P state of helium for incident electron energy of 80 eV and extended scattering angles. Sections 3.5.1,

4.12, 5.5 and Chap. 6 describe the apparatus, the experimental technique, the results and discussion and the concluding remarks, respectively.

## 1.6 Polarization Correlation Measurements on the $3^3$P State of Helium

The first angular-correlation measurements were reported by Eminyan et al. (1973, 1974, 1975). The basic theory of the electron-photon coincidence method was largely developed by Macek and Jaeck (1971), Fano and Macek (1973) and Blum and Kleinpoppen (1975). Experimental and theoretical work has been reviewed by Blum and Kleinpoppen (1979), Blum and Slevin (1984), Andersen et al. (1986) and Fursa and Bray (1997).

A number of experimental reasons like long lifetimes and considerable depolarization of the emitted light through fine-structure coupling precluded measurements on light systems to be extended to excited states with different multiplicity from the ground state, for example to triplet states in systems with a singlet ground state. As long as LS coupling holds, these states can be excited to electron exchange processes so that the pure exchange amplitudes can be measured and compared with theory. Previous studies of the $1\,^1$S $-\,3\,^3$P excitation in helium have been reported among others by Humprey et al. (1987), Donnelly et al. (1988), Donnelly and Crowe (1989), Batelaan et al. (1990) using unpolarized electron beam and Ding Hai-Bing et al. (2005) using polarized electron beam. Complementary theoretical studies have been reported by Cartwright and Csanak (1986) using first-order many-body theory (FOMBT) at energies in the range 30–500 eV, Bartschat and Madison (1988) at 40, 60, and 80 eV using the distorted-wave Born approximation (DWBA), Fon et al. (1990, 1991, 1993, 1995) using the 19- and 29-state R-matrix approach at energies up to 31.2 eV and Fursa and Bray (1995, 1997) using the convergent close-coupling (CCC) method, only at 30 eV.

We report here details of complete polarization correlation measurements at an extended range of scattering angles. The electron impact excitation of the $3\,^3$P state of helium is accompanied by the emission of light of the wavelength 388.9 nm,

$$e + \text{He}\left(1\,^1\text{S}\right) \rightarrow \text{He}^*\left(3\,^3\text{P}\right) + e \rightarrow \text{He}^*\left(3\,^3\text{P}_{0,1,2}\right) \rightarrow \text{He}^*\left(2\,^3\text{S}_1\right) + h\nu.$$

As shown, the process can be divided with good approximation into three independent stages: collisional excitation process ($\approx 10^{-16}$s), fine-structure coupling into the three $3\,^3$P states (0.1 ns) and decay ($\approx 100$ ns). The fine-structure coupling causes a reduction of the polarization of the decay light, but since the coupling is complete by the time the decay is detected, a correction can be applied to take account of the depolarization. However, the accuracy of the corrected polarization results is reduced by the depolarization, so that the measurements are extremely time consuming.

Sections 3.5.2, 4.13, 5.6 and Chap. 6 describe the apparatus, experimental technique, results and discussion and the concluding remarks, respectively.

# Chapter 2
# Theoretical Approaches

**Abstract** Theoretical approaches are described for the measurements of double differential cross sections (DDCS) and partial double differential cross sections (PDDCS) for the ionization of helium, argon, krypton and xenon atoms and hydrogen, sulphur dioxide and sulphur hexafluoride molecules by electron impact. Also the theoretical basis of the electron impact excitation of the hydrogen-like spin-polarized atoms of sodium and potassium and the helium and helium-like atoms of calcium and strontium is discussed.

**Keywords** Atoms · Excitation · Ionization · Molecules · Theoretical models

## 2.1 Ionization Cross Sections of Atomic Gases by Electron Impact

A complete theoretical treatment of the electron impact ionization remains a fundamental problem of atomic physics (see, e.g. Jacobowics and Moores 1983; Reid R. H. G. in *Photon and Electron Collisions with Atoms and Molecules* p. 37, edited by P. G. Burke and C. J. Joachain, Plenum Press, New York 1997). Reviews of the theory are given by many authors (see, e.g. Rudge 1968; Peterkop 1977; Reid 1997). The source of the physical problems is the long-range nature of the Coulomb potential which ensures that the two continuum electrons interact with the residual ion and each other until they are well apart. A complete treatment of the ionization process, therefore, requires a full solution of the three body problem in the asymptotic region. In restrictive calculations it is usual to make several approximations about target states, incident particle waveforms and to neglect correlation between the continuum electrons.

Atomic ionization can be expressed in terms of the various cross sections, for example total cross section, partial cross sections and differential cross sections, of various degrees of electron spin and correlation effects (see American Institute of

A. Chaudhry and H. Kleinpoppen, *Analysis of Excitation and Ionization of Atoms and Molecules by Electron Impact*, Springer Series on Atomic, Optical, and Plasma Physics 60, DOI 10.1007/978-1-4419-6947-7_2, © Springer Science+Business Media, LLC 2011

Physics Conference Proceedings, p. 697, G. F. Hanne, L. Malegat, H. Schmidt-Boecking, edits., New York, 2003). A restrictive representation of ionization is provided by the determination of the energy and momentum of all particles involved in the collision process. The triple differential cross section (TDCS), in the case of single ionization under electron impact, thus defined is given by the following (Ehrhardt et al. 1972):

$$\frac{\mathrm{d}^3\sigma}{\mathrm{d}E\mathrm{d}\Omega_A\mathrm{d}\Omega_B} = f(E_0, E_A, \theta_A, \theta_B, \phi_B)$$

where $E_O$ is the incident electron energy, $E_A$ the energy of one of the outgoing electrons and $\theta_A$ the angle with the incident electron direction while $\Phi_B$ and $\theta_B$ define the direction of the secondary electron. The double differential cross section (DDCS), differential in the energy of the scattered electron and the direction of one of the outgoing electrons, can be obtained by integrating the TDCS over the direction of one of the outgoing electrons. The integration of TDCS over the direction of both outgoing electrons yields the single differential cross section (SDCS), which is differential in the energy (or the angle) of the secondary electron. Further integration of the SDCS over the energy (or the angle) of the secondary electron yields the total ionization cross section.

### 2.1.1   Double Differential Cross Sections of Ionization

There are several approximations available for the calculation of the DDCS for the ionization of atoms. The First Born Approximation (Massey and Mohr 1933; Rudd et al. 1966; Oldham 1965, 1967; Tahira and Oda 1973), the Plane-Wave Born Approximation (Wetzel 1933; Glassgold and Ialongo 1968, 1969; Vriens 1970; Cooper and Kolbenstredt 1972; Tahira and Oda 1973; Kim and Inokuti 1973; Bell and Kingston 1975; Manson et al. 1975) and the Binary Encounter Theory (Vriens 1969; Bonsen and Vriens 1970; Tahira and Oda 1973) are considered to be practical methods (Tahira and Oda 1973) for the calculation of DDCS. These are discussed here briefly.

### 2.1.2   First Born Approximation

The TDCS for the ionization of hydrogen atom by electron impact has been given in the First Born Approximation by Massey and Mohr (1933), Mott and Massey (1965), Massey et al. (1969), and Landau and Lifschitz (1965). The sum of DDCS within this approximation can be represented as the DDCS for scattered electrons and that for the ejected electrons (Tahira and Oda 1973), where for mathematical simplicity either one of the two outgoing electrons is called "ejected" and the other is called "scattered." In fact, it is not possible to differentiate between the scattered and the ejected electron.[1] The DDCS for ejected electrons is obtained by integrating the TDCS over the direction of scattered electrons while the DDCS for the scattered electrons is calculated by integrating the TDCS over the direction of the ejected electrons. The values of the DDCS for the hydrogen atom have been calculated by Tahira and Oda (1973). The DDCS for other atoms can be obtained from that of hydrogen atom using the scaling methods of Rudd et al. (1966) or Tahira and Oda (1973). Bonsen and Vriens (1970) have shown in the case of proton impact, that the scaling of hydrogenic cross sections for helium on the expectation value for the kinetic energy of the atomic electrons (19.49 eV) leads to cross-section values that are in much closer agreement with the more accurate Hartree-Fock ones than the scaling on the ionization potential $U$(24.58 eV). The scaling procedures using 39.49 and 24.58 eV values are equivalent to the use of $z = 1.704$ and 1.344, respectively, in the scaling equations of Tahira and Oda (1973). Here $z$ is the effective nuclear charge.

Bell and Kingston (1975) have calculated the values of the DDCS for helium by electron impact at energies between 200 eV and 2,000 eV, using the First Born Approximation. Their conclusions, after comparison with the experimental results, are that

1. The Born approximation is unreliable below 200 eV incident electron energy.
2. At 500 eV incident electron energy there is a good agreement.
3. At an incident electron energy of 2,000 eV, the only serious disagreement between theory and experiment is in the forward scattering direction and for the slow-ejected electrons.

### 2.1.3 Plane-Wave Born Approximation

The triple differential cross section in this approximation is given in the atomic units by Glassgold and Ialongo (1968, 1969) and Vriens (1970)

$$\frac{d^3\sigma}{dEd\Omega_e d\Omega_S} = \frac{4k_e k_S}{k_i} \left[\frac{1}{q^4} + \frac{1}{S^4} - \frac{1}{q^2 S^2}\right] \times |\Phi_i (k_S + K_e - k_i)|^2 \delta (k_i^2 - k_S^2 - k_e^2 - 2U),$$

(2.1)

where $S$ is the magnitude of the exchange momentum transfer vector, $S = (k_i - k_e)$, $q$ is the magnitude of the direct momentum transfer vector, $q = k_i - k_S$, and $\Phi_i (k)$ is the initial state wave function of the target atom in momentum space. In the

---

[1] By applying electron spin effects in electron atom scattering it may be possible to distinguish between the scattered and ejected electron, for example in using spin-polarized electrons e($\downarrow$) scattered by spin-polarized hydrogen or alkali-atoms A($\uparrow$), e($\downarrow$) + A($\uparrow$) $\rightarrow$ A$^+$ + e($\downarrow$) + e($\uparrow$). (Kleinpoppen in: "Constituents of Matter", p. 314, de Gruyter, W. Reith, editor), New York, Berlin (1987).

particular case of helium when the hydrogenic wave function is assumed for the initial state, the expression for $|\Phi_i(k)|^2$ is given by the following equation (Sneddon 1951):

$$|\Phi_i(k)|^2 = \frac{8Z^5}{\pi^2} \times \frac{1}{(k^2 + Z^2)^4} \tag{2.2}$$

The direct term of (2.1) is identical with that obtained by Wetzel (1933), if one treats hydrogenic atoms and takes account only of the $e^2/r_{12}$ term in the perturbation, $r_{12}$ being the distance between the colliding electron and the atomic electron.

When $k_S$ is taken as the momentum of the detected electron, the DDCS is obtained by integrating

$$\frac{d^3\sigma}{dEd\Omega_A d\Omega_B}$$

over the direction of $k_e$, and the DDCS can be written (Tahira and Oda 1973) in units of $(a_o^2/2R_Y)$ as follows:

$$\frac{d^2\sigma}{dE_S d\Omega_S} = \frac{32Z^5}{\pi^2} \frac{k_S k_e}{k_i} n\{\sigma_D + \sigma_{EX} + \sigma_i\} \tag{2.3}$$

where $\sigma_D$, $\sigma_{Ex}$ and $\sigma_i$ are the direct, exchange and interference terms, respectively, $z$ is the atomic number and $n$ is the number of atomic electrons ($n = 2$ for helium). The expressions for $\sigma_D$, $\sigma_{EX}$ and $\sigma_i$ are given by Tahira and Oda (1973).

At intermediate and higher incident electron energies the theoretical PWBA calculations for DDCS agree fairly well with the experimental measurements (Tahira and Oda 1973).

Manson et al. (1975) have given calculations for DDCS, based on the Born approximation with Hartree-Slater (HS) initial discrete and final continuum wave functions for helium. Their calculations show a good agreement with the experimental values except at 30 and 150° secondary electron ejection angles.

### 2.1.4 Binary Encounter Theory

In the Binary Encounter Theory (BET) (Vriens 1969), an incident electron is supposed to interact with only one of the atomic electrons at a time and the cross sections for the electron-atom collisions are obtained by integrating the cross sections for the binary encounter collisions between incident and atomic electrons over the momentum distribution of the atomic electrons. The DDCS in this approximation is given in terms either of the energy of ejected electrons $E_e$ or the energy of scattered electrons $E_S$. The direct, exchange and interference terms are taken into account. The calculation was carried out in terms of $E_e$.

The DDCS is given in atomic units by the following equation (Tahira and Oda 1973)

$$\frac{d^2\sigma}{dE_e d\Omega_e} = n \int_{k_{min}}^{k_{max}} \{\sigma_D(k) + \sigma_{EX}(k) - \frac{1}{2}\sigma_i(k)\} f(k)dk, \tag{2.4}$$

where $\sigma_D(k)$, $\sigma_{EX}(k)$ and $\sigma_i(k)$ are the direct, exchange and interference terms, respectively, $n$ is the number of atomic electrons, $k_{min}$ and $k_{max}$ are the lower and upper bounds (Bonsen and Vriens 1970), respectively, for the momentum of atomic electrons contributing to the differential cross sections and $f(k)$ is the momentum distribution of atomic electrons. The first term of (2.4), the direct term, was first formulated by Bonsen and Vriens (1970) for the case of proton impact. The second and third terms, in this equation, were derived by Tahira and Oda (1973).

At relatively low incident electron energies and detection angles between 30 and 90° the agreement between the theoretical and the experimental results is quite well (Tahira and Oda 1973). As the binary encounter theory (BET) does not include phase shift effects, it cannot be expected to represent the DDCS properly at large angles (Bonsen and Vriens 1970).

## 2.1.5  The Inner Shell Ionization

The ionization of inner shells can be affected in several ways. The expression for the ionization cross section under electron impact as obtained by Mott and Massey (1965) is as follows:

$$Q_{ni} = \frac{2\pi e^4 Z_{nl}}{mvE_{nl}} b_{nl} \log_e \frac{2mv^2}{B_{nl}}, \tag{2.5}$$

where $Q_{nl}$ is the cross section for the $(nl)$th shell, $e$ and $m$ is the electronic charge and mass, respectively, $v$ is the velocity of the incident electron, $b_{nl}$ and $B_{nl}$ are constants, and $z_{nl}$ is the number of electrons in the $(nl)$th shell. For $K$ shell Burhop (1940) has given values of 0.35 for $b_{nl}$ and 1.65 $E_{nl}$ for $B_{nl}$, where $E_{nl}$ is the ionization potential of the shell. The logarithm in the (2.5) can be written as $\log_e (4E/1.65 E_k)$ where $E$ is equal to the kinetic energy of the incident electron and $E_k$ is the binding energy of the $K$ shell. The Bethe-Bloch energy-loss equation given by Segré (1959) has the logarithm term in the form $\log_e (2E/E_k)$ and there is some uncertainty in the details of this part of expression which is important in the region where $E$ does not greatly exceed $E_k$.

Worthington and Tomlin (1956) have derived an empirical formula for $B_{nl}$ which approaches 1.65 $E_k$ for large excitation voltages, but which allows $B_{nl}$ to approach 4 $E_k$ for excitation voltages just exceeding the excitation limit ($U > 1$). Their expression for $B_{nl}$ is as follows:

$$B_{nl} = [1.65 + 235 \exp(1 - U) E_k \tag{2.6}$$

where $U$ is the excitation ratio $E/E_k$. Equation (2.5) becomes

$$Q_{nl} = \frac{2\pi e^4 b_{nl}}{UE_k^2} \log_e \frac{4UE_k}{B_{nl}}$$
(2.7)

Equations (2.5) and (2.7) show that the expression $Q_k E_k^2$ is the same function of $U$ for all elements and that, $Q_k E_k^2$ for $U > 1$, can be written in a simple form (Worthington and Tomlin 1956) as follows:

$$Q_k E_k^2 = \frac{2\pi e^4}{U} (0.7)\log_e U.$$
(2.8)

If $E_k$ is expressed in electron volts this equation takes the form

$$Q_k E_k^2 = 2\pi e^4 \frac{\log_e U}{U} \times 6.3 \times 10^4 \quad \text{(e in e.s. units)}$$
$$= 9.12 \times 10^{-14} \frac{\log_e U}{U} \quad \quad (eV)^2\,(cm)^2.$$
(2.9)

The cross sections for the heavier elements are in fact considerably greater than those expected from equations (2.8) and (2.9). Some experimental data has been shown to be in agreement with the relativistic calculations of Arthurs and Moise-witch (1958). A discussion of relativistic cross sections for $K$-shell ionization is given by Perlman (1960).

Hippler and Jitschin (1982) have calculated $K$-shell ionization cross sections for light atoms using the plane-wave Born approximation (PWBA) and Ochkur approximation. Their inclusion of exchange effects has considerably improved the agreement with experimental results. Use of be distorted-wave Born approximation (DWBA) (Madison and Shelton 1973) and Coulomb-exchange method (Moors et al. 1980) may resolve the remaining inadequacies (Hippler and Jitschin 1982).

## 2.1.6   Electron Vacancy Transitions of Atoms

The various types of electron vacancy transitions of inner shells of atoms are schematically illustrated and explained in Fig. 2.1, it demonstrates the emission of photons and electrons from the sub-shells with vacancies. The emission of an electron is an "auto-ionization" decay which can be distinguished as Auger electrons, Coster-Kronig electrons and Super Coster-Kronig electrons.

### 2.1.6.1   Auger Effect and Auger Transitions

The Auger effect (also see Sect. 3.1.10) was discovered by P. Auger (1925). Wentzel (1927) gave a non-relativistic theory for Auger transitions which was

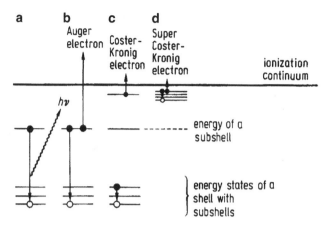

**Fig. 2.1** The various decay types of an electron hole (*open circle*) of inner shells of atoms: (**a**) X-ray transition (*hv*), (**b**) Auger transition (*closed circle*), (**c**) Coster-Kronig transition (*closed circle*), (**d**) super Coster-Kronig transition (*closed circle*)

reviewed by Burhop (1952). An Auger effect and its transitions can be induced by innershell ionization.

Let us assume that a projectile P (i.e. a photon or an atomic particle such as an electron, ion, ...) induces the ionization of an atomic inner shell, the Auger effect may then be represented by the following reaction processes:

$$(1) \qquad P + A \rightarrow P + A^{+} + e(E_1) \rightarrow P + A^{++} + e(E_1) + e(E_{Auger}),$$

$$(2) \qquad P + A \rightarrow P + A^{++} + e(E_1) + e(E_{Auger}),$$

where A represents the target atom, $e(E_1)$ is the knocked-out electron with the energy $E_1$ and $e(E_{Auger})$ is the Auger electron with the sharp energy $E_{Auger}$. It has been found that the direct double-ionization process (2), as a one-step process, has a much lower probability than the two-step process (1). It is common to characterize the Auger electrons as Auger transitions in analogy to the characteristic X-ray lines. The notations K-LL or briefly KLL means that initially an electron is knocked out of the $K$-shell and subsequently two electron holes in the $L$-shell are "produced" by the Auger effect. Correspondingly KMM, LMN, Auger transitions are possible, whereby the sub-shells can be classified in addition by $KL_IL_{II}$, $KL_IL_{III}$ or by $KL_1L_2$, $KL_1L_3$. The double holes are also characterized by the coupling mechanism such as $LS, jj$ and the intermediate coupling.

The Auger, Coster-Kronig and Super Coster-Kronig transitions have sharp energies which result from the energy differences of the inner shells and the sub-shells. Accordingly, the detection of Auger and Coster-Kronig transitions can be carried out by an electron energy analyzer. A frequently applied electron energy analyzer is the cylindrical 127° analyzer shown schematically in Fig. 2.2a. Two

**Fig. 2.2** Schematic arrangement of a spectrometer for the detection of Auger electrons including a collision chamber and a 127° electron energy analyzer. The Auger electrons are produced by electrons impinging on the atomic target in the centre of the collision chamber

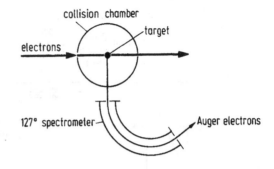

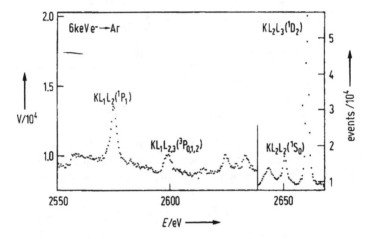

**Fig. 2.3** KLL Auger-electron spectrum of argon atoms; the peaks not specifically assigned are Auger satellite lines. $N$ is the number of the counted events (after Gräf 1985)

cylindrically shaped metal plates are kept at positive and negative potentials, producing a radial electric field in the plane of the Fig. 2.2a, the electric force $eE$ then keeps the electron on the circular trajectories if the centrifugal force is compensated, i.e. $eE = mv^2/r$. It can be shown that electrons passing through the entrance slit of the analyzer within a small angle will be focused on to the exit slit. Varying the electric field strength changes the velocity $v$ or the electron energy. Such a 127° energy analyzer (or other types of all analyzers for that matter) can be applied to detect Auger or Coster-Kronig electrons. Figures 2.3 and 2.4 show typical examples of Auger and Coster-Kronig spectra.

If the atom, in addition to the inner shell vacancy, is multiply ionized in the outer shells, the Auger spectrum contains only satellite lines. In the case when more than two electrons are involved in the transition, for example K-LLL or K-LLL$^*$ or KK-LLL, the transitions give rise to correlation satellite lines. Transitions such as K-LLL or K-LLL$^*$, where either two electrons are ejected simultaneously (Carlson and Krause 1965, 1966; Aberg 1975) or one electron is ejected and another electron is in an excited state (Mehlhorn 1976), are known as double Auger transitions. In

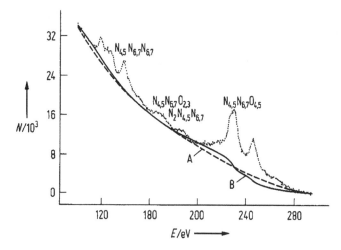

**Fig. 2.4** NNO Coster-Kronig and NNN super Coster-Kronig lines of mercury atoms produced by excitation with 3-keV electrons. The curves A and B describe theoretically the background signals of the electrons. $N$ is the number of events counted (after Aksela and Aksela 1983)

the K-LLL transition both ejected electrons share the transition energy leading to a continuous energy distribution:

$$E_{Auger1} + E_{Auger2} = E(k) - E(LLL)$$

in the K-LLL$^*$ transition the energy of the ejected electron is smaller than it would otherwise be by an amount equal to the excitation energy of the other excited electron and is given by the following:

$$E_{Auger} = E(k) - E(LLL) - E_{excitation}.$$

Transitions such as KK-LLL, three-electron Auger transitions in an atom with two vacancies, have been found, by Afrosimov (1976) and Shergin and Gordeev (1977), in heavy ion-atom collisions where the formation of two $K$ vacancies has a much larger cross section.

### 2.1.6.2 Coster-Kronig Transitions

At a transition where one of the final vacancies occurs in the same main shell but different sub-shell is referred to as a Coster-Kronig (CK) transition (see also Sect. 3.1.10), for example $L_1L_{23}M$. The energy of the Coster-Kronig electrons is correspondingly smaller and in some cases Coster-Kronig transitions are forbidden. On the other hand, Coster-Kronig transitions may sometime have a much larger transition probability ($\approx 10$ times) than the competing Auger transitions (e.g. $L_1MM$).

If both final vacancies occur in the same main shell, their transitions are called Super Coster-Kronig (SCK) transitions. Because of the energy considerations, these transitions can only occur in atoms with atomic number $z$ within a certain range (McGuire 1974; Chen et al. 1976). Their occurrence changes the photo-electron spectra completely. For example, for $z < 54$ due to the strong decay probability of the 4p vacancy, the 4p photo-electron is completely diluted in the background and/or shifted several electron volts to lower energies (Wendin and Ohno 1976; Krause 1976).

### 2.1.7 Characteristic and Continuum X-Rays

Both characteristic and continuum X-rays were discovered by W. Röntgen in 1896. Characteristic X-rays can arise from the rearrangement of an electron in a state with given orbital and spin quantum numbers to another state with different quantum numbers (inner-shell excitation or ionization process). When a vacancy in an inner-shell is refilled, the atom changes to a state of lower energy and this excess energy may be released in two ways: either an X-ray photon may be emitted or alternatively a radiation-less transition may take place in which the available energy is used to release an electron from an outer shell.

To calculate the relative intensities of lines in allowed X-ray transitions we apply the "sum rule" which states that, for the lines comprising a multiplet, the total intensity of all lines proceeding from a common initial level or to a common final level is proportional to the statistical weight $(2J + 1)$ of that level. The relative intensity of some $K$ and $L$ series has been calculated by Beckman (1955).

The width of the lines of the characteristic X-ray spectrum has been examined as a function of atomic number and is found to exhibit several interesting features. It is well known that the energy width of a state and its lifetime are related by the Heisenberg uncertainty principle $\Gamma \tau = \hbar$, where $\Gamma$ is the width in energy units and $\tau$ the mean lifetime of the state. If $P$ is the probability per unit time of the transition, then one can write $\tau = P^{-1}$ and $\Gamma = \hbar P$. In the case of a transition from a state of inner-shell ionization, the probability of radiative and non-radiative (Auger) transitions may be written as $P_r$ and $P_n$, respectively, and the lifetime will thus be given by $(P_r + P_n)^{-1}$. The presence of competing processes thus reduces the lifetime of the state and must, therefore, through application of the uncertainty principle, increase the width of the state. The total width $\Gamma_t$ may be defined as the sum of two partial widths $\Gamma_r$ and $\Gamma_n$, where $\Gamma_r = \hbar P_r$ and $\Gamma_n = \hbar P_n$.

### 2.1.8 Close-Coupling Approach to Electron-Impact Ionization

The close-coupling techniques were developed in the early 1930s by Massey and Mohr (1932) who gave a general formalism for treating the discrete atomic transitions. The method consists in expanding the total wave function by using square-integrable states. Since the close-coupling equations yield stationary amplitudes upon variation in the expansion of the total wave function it is not surprising that they have been so successful (Bray 2002) in treating discrete transitions in the various collision systems. Bray and Fursa (1996) have suggested that the extension of the convergent close-coupling (CCC) method also yields accurate ionization amplitudes as long as sufficient computational resources are utilized in their evaluation. Stelbovics (1999), Bray et al. (2001), Bray (2002.), Bray et al. (2003, 2006), Colgan et al. (2009) and others have used the close-coupling formalism and have been successful to some extent in solving Coulomb three-body problems such as electron-hydrogen-atom collision and the electron impact single ionization of helium atom and their calculations in most cases agreed with the experimental values from recent literature.

### 2.1.9 Partial Double-Differential Cross Section for Ionization

Double-differential cross section for $n$-fold ionization or partial double-differential cross section, DDCS($n+$), can be approximated to the double-differential cross section (DDCS) as follows:

$$\mathrm{DDCS} = \sum_n \mathrm{DDCS}\,(n+),$$

where $n$ is the charge state of ionization.

The values of DDCS mainly reflect the $M$-shell DDCS since $K$- and $L$-shell DDCS are comparatively small (Hippler 1984c). To obtain DDCS and DDCS($n+$) for $K$- and $L$-shell one has to perform a coincidence experiment between

(a) Ejected electrons and Auger electrons or characteristic X-rays
(b) Ions and Auger electrons or characteristic X-rays

following the decay of specific inner-shell vacancies, to effectively suppress the detection of ejected electrons from other shells.

The DDCS($n+$) can give very useful information about the multiple ionization in the collision process. Unfortunately, not much theoretical work exists in the literature in this regard and the only experimental investigations for the measurement of DDCS($n+$) are as follows:

1. For the electron-argon-atom collision process by Hippler et al. (1984b)
2. For proton collisions with helium, neon and argon gas atoms by Hippler et al. (1984a)
3. For electron-rare-gas-atom collisions by Chaudhry et al. (1986).

For the present measurements of DDCS($n+$) for the rare gas atoms the apparatus used is described in the Sects. 3.1.1–3.1.12.6, the experimental techniques are given in the Sections 4.1–4.3, the results and discussions are recorded in the Sects. 5.1–5.1.6 and the concluding remarks are given in the Chap. 6.

## 2.2 The Ionization of Hydrogen, Sulphur Dioxide and Sulphur Hexafluoride

Electron impact can remove an electron from a molecule thereby producing a single ionization of the molecule without any dissociation of the molecule (non-dissociative ionization) or can break up the molecule and also produce ionization of one or more fragments of the molecule (dissociative ionization). The dissociative ionization of a molecule by electron impact may occur via different reaction channels involving direct or sequential ionization processes. Each of these processes produces different types of ions and sometimes more than one process can give rise to the same ion.

The minimum energy ($U_{min}$), required for an electronic transition leading to the dissociation or dissociative ionization resulting in atoms and/or ions having some relative kinetic energy, is given by the following equation (Massey et al. 1969):

$$U_{min} = U_A + U_B + D_{AB} + W_{min} \qquad (2.10)$$

where $U_A$ and $U_B$ are the excitation energies of the two atoms of the molecule AB; $D_{AB}$ is the dissociation energy and $W_{min}$ is the minimum energy of the relative motion of the resulting fragments of the molecule. If under an electron impact the molecule AB breaks into an atom A of mass $M_A$, and an ion $B^+$ of mass $M_B$, and the ion $B^+$ has a measured value of the kinetic energy $W^+$ then the total kinetic energy $W$ of the atom A and the ion $B^+$ is given by the conservation of momentum as

$$W = (1 + M_B/M_A)W^+ \qquad (2.11)$$

In a dissociative ionization of hydrogen by electron impact reaction

$$e + H_2 \rightarrow H + H^+ + 2e,$$

the appearance potential (AP) of the proton with zero kinetic energy, both products being in their ground states, is equal to $U_{min}$ (the minimum energy required for this reaction), given by the sum of the dissociation energy($D_{AB}$) of $H_2$ (4.5 eV) and the ionization energy ($U_B$) of H atom (13.6 eV) is $\approx$18 eV. For an ionization process

$$e + H_2 \rightarrow H^+ + H^+ + 3e,$$

the minimum energy ($U_{min}$) ionization energy of H to the $U_{min}$ for the reaction($e + H_2 \rightarrow H + H^+ + 2e$).

If a target in its initial ground electronic state $\Psi_i$ is bombarded by an electron of energy $E_o$, which exceeds the appearance potential (AP) of the target ion, then the

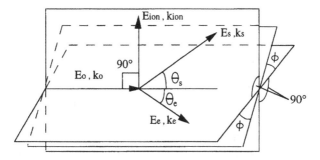

**Fig. 2.5** A schematic diagram of the kinematics of an ionizing electron collision with an atom. The energy and direction of the ejected and scattered electron and the produced ion are illustrated

ionization may occur and a scattered electron of energy $E_s$ and an ejected electron $E_e$ (see Fig. 2.5) leave the collision region resulting in making angles $\theta_s$ and $\theta_e$, respectively, with the direction of the incident electron. This leaves the produced ion in a final state $\Psi_f$. In a collision process resulting in ionization, the incident electron, the two outgoing electrons and the produced ion may not be moving in the same plane. Figure 2.5 shows the collision kinematics in the present work. Here $\phi$ is the angle between the planes $(k_o, k_e)$ and $(k_o, k_s)$ and an angle of $90^\circ$ between the planes $(k_o, k_e)$ and $(k_o, k_{ion})$, where $k_o$, $k_e$ $k_s$ and $k_{ion}$ are the momenta of the incident, ejected and scattered electrons and the produced ion, respectively. If the ion is left in an excited state of energy $E_{ex}$ then energy conservation requires that

$$E_o = E_e + E_s + AP + E_{ex} + \delta E \qquad (2.12)$$

where $\delta E$ is the kinetic energy imparted to the target in the collision process and is $\leq 4 \, (m_e/M) \, E_o$ (Van Brunt and Keifer 1970), which is very small and can be neglected here. If the ion is in its ground state, that is $E_{ex} = 0$, then

$$E_o = E_e + E_s + AP \qquad (2.13)$$

In the present work, the coincidences are measured between the ejected-electron in plane $(k_o, k_e)$ and the produced ion in the plane $(k_o, k_{ion})$ while the scattered electron is not taken into account. In this case, the remaining energy $(E_o - E_e - AP)$ gives the sum of the scattered electron energy and molecular ion excitation energy (including electronic, vibrational and rotational energies) while if the ion is in the ground state, the scattered electron carries the remaining energy.

For the measurement of the ionization cross sections, dissociative and non-dissociative, of molecules different methods have been employed and data exists for total ionization cross sections (Orient and Srivastava 1987; Rapp and Englander-Golden 1965; Schram et al. 1965; Jain and Khare 1967), partial ionization cross sections including dissociation (Kieffer and Dunn 1966; Barton and Von Engel 1970; Adamczyk et al. 1972; Kim et al. 1981; Shah and Gilbody 1982), double ionization cross sections (Peresse and Tuffin 1967; Halas and Adamczk 1972/73; Crowe and McConkey 1973a, b; Mark 1975; Hille and Mark 1978; Edwards et al. 1989) and for multiple ionization cross sections (Dorman and Morrison 1961; Schram et al. 1965; Ziezel 1967). For more information about the ionization process, reference should be made to the work on differential ionization cross sections including that for singly differential cross sections (Ehrhardt et al. 1969; Ehrhardt et al. 1971; Omidvar et al. 1972; Kim and Inokuti 1973; Cheng et al. 1989), doubly differential ionization cross sections (Peterson et al. 1971; Opal et al. 1972; Peterson et al. 1972; Tahira and Oda 1973; Dubois and Rudd 1978; Shyn and Sharp 1991; Rudd 1991; Rudd et al. 1993), triply differential ionization cross sections (Ehrhardt et al. 1969; Vriens 1970; Ehrhardt et al. 1972a; Camilloni et al. 1972; Manson et al. 1975; Vucic et al. 1987; Ray and Roy 1988; Cherid et al. 1989) or even fourfold and fivefold differential ionization cross sections (Lahmam-Bennani et al. 1989, 1991; Hafid et al. 1993; Hanssen et al. 1994).

The apparatus for the present investigations is described in the Sect. 3.2, the experimental arrangement is given in the Sect. 4.4, the measurements are shown in the Sects. 5.2.1–5.2.3 and the concluding remarks are given in the Chap. 6.

## 2.3   Excitation of Spin-Polarized Sodium and Potassium Atoms by Electron Impact

A theoretical treatment of the excitation of spin-polarized sodium and potassium atoms by electron impact is given by Jitschin et al. (1984). The atomic beam in this treatment, is considered as a mixture of states such that the density matrix which represents this mixture can be expanded in a series of state multipoles $\langle T^{\dagger}{}_{KQ} \rangle$ of different rank $K_r$, that is monopoles, dipoles, etc. (Blum 1981). The direction of the magnetic field, which is also the direction of the spin of the outer electron of the atom, is used, in the following discussion, as the quantization axis (or the polarization frame).

### 2.3.1   Preparation of the State-Selected Na and K Atomic Beams

The sodium and potassium atomic beams have been polarized using a hexapole magnet having achieved a polarization of approximately 21% in a low magnetic field. The atomic beam, in a low magnetic field, can be described as an incoherent set of atoms being in different $FM_F$ hyperfine states and the density matrix of such a beam is diagonal. The elements of the density matrix are the occupation numbers $W$ $(FM_F)$. The density matrix can be expanded into a series of state multipoles $\langle T(F)^{\dagger}{}_{K_F Q_F} \rangle$ which are actually the set of $(2F + 1)^2$ multipole operators pertaining to the orientation of particles having angular momentum $F$. Each operator $\langle T(F)_{K_F Q_F}{}^{t} \rangle$ is represented by a $(2F + 1) \times (2F + 1)$ matrix. The multipole operators are chosen in a tensor form $\langle T(F)_{K_F Q_F}\rangle$ so that they transform under coordinate rotations like the spherical harmonics (Drukarev 1987)

$$Y_{KQ}, Q = -K \ldots + K \text{ and } K = 0, \ldots 2F.$$

The state multipole $\langle T(F)^\dagger{}_{K_F Q_F}\rangle$ is given by the following equation:

$$\langle T(F)^t{}_{K_F Q_F}\rangle = \sum_{M_F M_F} (-1)^{F-M_F}(FM_{F'}F - M_F|K_F Q_F)\delta_{M_F M'_F}W(FM_F), \qquad (2.14)$$

where $(|)$ is the Clebsch-Gorden coefficient.

Using the electronic and the nuclear spin parameters the collision representation by the uncoupled state multipoles is $\langle T(S)^\dagger{}_{K_S Q_S} \times T(I)^\dagger{}_{K_I Q_I}\rangle$ (Blum 1981) which can be written as follows:

$$\begin{aligned}
&\langle T(S)^\dagger{}_{K_S Q_S} \times T(I)^\dagger{}_{K_I Q_I}\rangle \\
&= \sum_{FK_F Q_F} (2F+1)[(2K_S+1)(2K_I+1)]^{1/2}(K_S Q_{S'}K_I Q_I|K_F Q_F) \\
&\times \begin{vmatrix} K_S & K_I & K_F \\ S & I & F \\ S & I & F \end{vmatrix} \langle T(F)^\dagger{}_{K_F Q_F}\rangle
\end{aligned}$$

$$(2.15)$$

where $\begin{vmatrix} \cdot & \cdot & \cdot \\ \cdot & \cdot & \cdot \\ \cdot & \cdot & \cdot \end{vmatrix}$ is a 9-j symbol and $I$ is the nuclear spin.

Further calculations show that the electronic spin polarization $P_S$ and the nuclear vector polarization $P_I$ can be expressed by these following multipoles:

$$P_S = <T(S)^\dagger{}_{10} \times T(I)^\dagger{}_{00}><T(S)^\dagger{}_{00} \times T(I)^\dagger{}_{00}>^{-1} \qquad (2.16)$$

$$P_I = (5/9)^{1/2}<T(S)^\dagger{}_{00} \times T(S)^\dagger{}_{10}><T(S)^\dagger{}_{00} \times T(I)^\dagger{}_{00}>^{-1}. \qquad (2.17)$$

For the ground state of sodium and potassium atoms the occupation number $W(FM_F)$ is given by (14),

$$W(FM_F) = (1/8)(1 \pm s) \qquad (2.18)$$

where s is the selectivity of the hexapole magnet. The negative sign $(-)$ is applicable, in (2.18), for $F = 1$ and $M_F = -1, 0, +1$ and also for $F = 2$ and $M_F = -2$, while the positive sign $(+)$ is applicable for $F = 2$ and $M_F = -1, 0, +1$ and $+2$. Assuming that the selectivity number s of the hexapole magnet is 0.7 (Hils et al. 1981), Table 2.1 gives the occupation numbers $W(FM_F)$ for sodium and potassium atoms.

**Table 2.1** Values of W(FM (subscript 'F')) for the atoms of sodium and potassium

| State | | Magnet state selection W(FM$_F$) of sodium and potassium atoms |
|---|---|---|
| $F$ | $M_F$ | |
| 1 | $-1$ | 0.0375 |
| 1 | 0 | 0.0375 |
| 1 | $+1$ | 0.0375 |
| 2 | $-2$ | 0.0375 |
| 2 | $-1$ | 0.2125 |
| 2 | 0 | 0.2125 |
| 2 | $+1$ | 0.2125 |
| 2 | $+2$ | 0.2125 |

### 2.3.2   The Collision Induced S-P Excitation

As suggested by Blum and Kleinpoppen (1979) we consider that in the excitation process all the angular momenta are decoupled. In these experiments the scattering plane is defined by the direction of the incoming electrons and the outgoing electrons. Figure 2.6 shows the angular shape of atoms in the ground state (or S state) which is isotropic and the excited states (or P states) which are anisotropic.

The collision experiments where the coincidence technique is not used and all the emitted photons are counted regardless of the scattered electron direction, so that the observed quantity is integrated over all the scattering angles of the electrons, the differential cross sections cannot be measured but the total cross sections, or integral cross sections, for the excitation of 3P states are measured (Moores and Norcross 1972). The total cross section $Q_M$ (by putting $M_1 = M$) is given by the following equation:

$$Q_M = 1/2(\mathrm{D}_M + \mathrm{E}_M + \mathrm{I}_M), \tag{2.19}$$

where

$$\mathrm{D}_M = (k_\mathrm{f}/k_\mathrm{i}) \int |f_M|^2 \mathrm{d}\Omega \tag{2.20a}$$

$$\mathrm{E}_M = (k_\mathrm{f}/k_\mathrm{i}) \int |g_M|^2 \mathrm{d}\Omega \tag{2.20b}$$

$$\mathrm{I}_M = (k_\mathrm{f}/k_\mathrm{i}) \int |f_M - g_M|^2 \mathrm{d}\Omega \tag{2.20c}$$

In the (2.20) $k_\mathrm{f}^2$ and $k_\mathrm{i}^2$ are, respectively, the final and the initial electron energies, $f_M$ and $g_M$ are the direct and the exchange amplitudes and $\Omega$ is the solid angle.

In an electron-atom scattering process for two unpolarized colliding partners, however, in which the electron beam is in the $OY$-direction, the atomic beam is in the $OX$-direction and the photon detector is in the $OZ$-direction, the experiment has a cylindrical symmetry about the $Y$-direction but if the atomic beam is

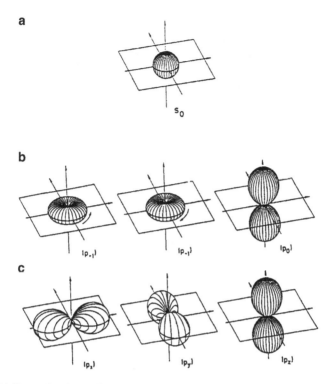

**Fig. 2.6** (**a**) Shows the charge cloud for the state S ($m = 0$); (**b**) shows the charge cloud for the state 3P with magnetic quantum number $M_1 = +1, -1$ and 0 in the atomic physics basis and (**c**) shows, the same as in (**b**), the molecular basis (Anderson 1988)

**Fig. 2.7** The geometry of the experiment at the interaction region

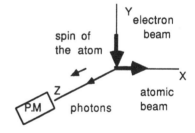

polarized along the Z-direction the collision geometry loses its axis of symmetry. In this case the polarization P of the atom defines a sense of rotation in the XY-plane and the geometry of the experiment would have only reflection symmetry in the XY-plane. In such a geometry $P_2$ (the linear polarization Stokes parameter corresponding to polarizer angle 45 and 135° with respect to Y-direction) is not necessarily zero.

Figure 2.7 shows the experimental symmetry in the collision region. For sodium atom, which in the excited state exhibits anisotropy that can be described by a state multipole of rank larger than zero, the only state multipoles which do not vanish due to symmetry are $\langle T^-(L)_{00}{}^\dagger\rangle$ and $\langle T^-(L)_{20}{}^\dagger\rangle$. The impact axis is considered to be the quantization axis as indicated in Fig. 2.7. The integrated multipoles are related to the excitation cross sections $Q_0$ and $Q_1$ of the magnetic sub-states $M_L = 0$ and $M_L = \pm 1$, respectively, by the following:

$$\langle \bar{T}(L)_{00}{}^\dagger\rangle = (1/3)^{1/2}(2Q_1 + Q_0) \qquad (2.21)$$

$$< \bar{T}(L)_{20}{}^\dagger > = (2/3)^{1/2}(Q_1 - Q_0) \qquad (2.22)$$

The rotation of the quantization axis through $90°$ would make it parallel to the spin polarization axis (or along Z-axis) where the photo-detector has been placed [instead of along the electron impact axis (Edmonds 1959)], the state multipoles in the new form can be obtained as follows:

$$\langle T(L)_{00}{}^\dagger\rangle = \langle \bar{T}(L)_{00}{}^\dagger\rangle \qquad (2.23a)$$

$$\langle T(L)_{2\pm2}{}^\dagger\rangle = (3/8)^{1/2}\langle \bar{T}(L)_{20}{}^\dagger\rangle \qquad (2.23b)$$

$$\langle T(L)_{2\pm1}{}^\dagger\rangle = 0 \qquad (2.23c)$$

$$\langle T(L)_{20}{}^\dagger\rangle = -1/2\langle T(L)_{20}{}^\dagger\rangle \qquad (2.23d)$$

As mentioned previously, for the interaction of electrons with polarized atoms, there are three channels of interaction namely direct (D), exchange (E) and mixed (I) which can be represented by the state multipoles $\langle T^D(L)_{K_LQ_L}{}^\dagger\rangle$, $\langle T^E(L)_{K_LQ_L}{}^\dagger\rangle$ and $\langle T^I(L)_{K_LQ_L}{}^\dagger\rangle$, respectively, and their values can be calculated if the corresponding cross sections for the magnetic substates are known.

In a collision experiment, the electron spin is conserved in the direct and the mixed channels while a spin flip occurs in the exchange channel (Burke and Schey 1962); Moores and Norcross 1972). Assuming that the nuclear spin is not affected by the electron atom collision process, the complete description of the excited state of an atom immediately after the collision in terms of the uncoupled state multipoles is given by Jitschin et al. (1984) as follows:

$$\langle T(L)_{K_LQ_L}{}^\dagger\rangle \times \langle T(S)_{K_SQ_S}{}^\dagger\rangle \times \langle T(I)_{K_IQ_I}{}^\dagger\rangle$$
$$= 1/2\langle T^D(L)_{K_LQ_L}{}^\dagger + T^I(L)_{K_LQ_L}{}^\dagger\rangle \langle T(S)_{K_SQ_S}{}^\dagger \times T(I)_{K_IQ_I}{}^\dagger\rangle$$
$$+ 1/2\langle T^E(L)_{K_LQ_L}{}^\dagger\rangle \langle < \pm T(S)_{K_SQ_S}{}^\dagger \times < T(I)_{K_IQ_I}{}^\dagger\rangle$$

$$= \frac{1}{2} < T^D(L)_{K_LQ_L}{}^\dagger \pm < T^E(L)_{K_LQ_L}{}^\dagger$$
$$+ \langle T^I(L)_{K_LQ_L}{}^\dagger\rangle \langle T(S)_{K_SQ_S}{}^\dagger\rangle \times \langle T(I)_{K_IQ_I}{}^\dagger\rangle. \qquad (2.24)$$

The state multipoles $\langle T(S)_{K_SQ_S}{}^\dagger \times T(I)_{K_IQ_I}{}^\dagger \rangle$ are the same as for the ground state and the $\pm$ signs apply for $K_S = 0$ and $K_S = 1$, respectively, for sodium.

### 2.3.3  Fine and Hyperfine Interaction of the Excited States

If the outer most electron of the alkali metal atoms rotates about the nucleus with an orbital angular momentum ($L$) and the intrinsic (spin) angular momentum ($S$) then the spin-orbit interaction between these causes fine structure in the atomic spectrum. The atomic nucleus also has an angular momentum ($I$), called nuclear spin, which causes a further splitting of the spectrum line, called the hyperfine structure. For sodium and potassium atoms the line splitting due to hyperfine structure is large compared to the natural level width of the spectrum lines (See Fig. 4.19 and 4.20). If the electron spin and the nuclear spin are both unpolarized then the effect of the fine and hyperfine structure coupling reduces the collisionally induced orientation and alignment by a certain factor which depends on the rank of the state multipoles. The state multipoles of different angular momentum are mixed but not the state multipoles of different rank and this results in the depolarization of the atomic beam by a certain factor depending on the depolarization coefficient $G_K$, where $K$ is the rank. Jitschin et al. (1984) have given the values of the depolarization coefficient for rank $K = 0$, 1 and 2 as follows:

$$G_0 = 1, \ G_1 = 0.390 \quad \text{and} \quad G_2 = 0.0982.$$

These values show that the orientation (rank 1) is less affected than the alignment (rank 2).

For the state-selected polarized sodium atoms, when the nucleus is aligned and oriented immediately after the excitation, the state multipoles of different rank can mix by the effect of fine and hyperfine coupling. The polarization properties of the emitted fluorescence light are determined by the spatial multipoles $\langle T(1)^\dagger_{k_1q_1} \rangle$ averaged over the decay time (the spatial state multipole is denoted by lower case indices). Jitschin et al. (1984), also, obtained the time-averaged perturbed multipoles

$$\langle T(L)^\dagger_{K_LQ_L} \times T(S)^\dagger_{K_SQ_S} \times T(I)^\dagger_{K_IQ_I} \rangle \langle T(1)^\dagger_{k_1q_1} \rangle$$
$$= [(2S+1)(2I+1)]^{1/2} \langle T(1)^\dagger_{k_1q_1} \times T(S)_{00}{}^\dagger \times T(I)^\dagger_{00} \rangle$$
$$= [(2S+1)(2+1)]^{1/2}$$

$$\sum_{\substack{K_L \ Q_L \\ K_S \ Q_S \\ K_I \ Q_I}} \langle T(L)^\dagger_{K_LQ_L} \times T(S)^\dagger_{K_SQ_S} \times T(I)^\dagger_{K_IQ_I} \rangle \, \underline{G} \begin{Bmatrix} Q_L & Q_S & Q_I & q_1 \\ Q_L & Q_S & Q_I & q_1 \end{Bmatrix} \tag{2.25}$$

and $\underline{G}$ is now given by the following:

$$
\underline{G}\begin{matrix} Q_L\ Q_S\ Q_I\ q_1 \\ Q_L\ Q_S\ Q_I\ q_1 \end{matrix} = \sum_{\substack{JF \\ J'F' \\ K_J}} \frac{[(2K_L+1)(2K_S+1)(2K_I+1)(2K_J+1)]^{1/2}}{1+(W_{JF}-W_{J'F'})^{2\pm2}}(2K_1+1)(2J+1)
$$

$$
\times (2J'+1)(2F+1)(2F'+1)(K_L Q_{L'} K_S Q_S | K_J Q_J + Q_S)
$$

$$
\times \begin{vmatrix} K_L & K_S & K_J \\ L & S & J' \\ L & S & J \end{vmatrix}\begin{vmatrix} K_J & K_I & K_1 \\ J' & I & F' \\ J & I & F \end{vmatrix}\begin{vmatrix} k_1 & 0 & k_1 \\ L & S & J' \\ L & S & J \end{vmatrix}\begin{vmatrix} k_1 & 0 & K_1 \\ J' & I & F' \\ J & I & F \end{vmatrix}.
$$

$$(2.26)$$

It is evident that, when the hyperfine coupling vanishes, the $\underline{G}$ coefficient with $K_I = Q_I = 0$ would only have any value thus reducing considerably the number of relevant coefficients. The coefficients $G$ and $\underline{G}$ contain both the angular momentum coupling factor of the excited state as well as the electronic and nuclear spin polarization factor of the excited atoms, immediately after the collision. It is, therefore, possible to investigate, in more detail, the influence of the electronic and nuclear polarization on the polarization of the fluorescent light.

### 2.3.4 Polarization of the Fluorescence Radiation

The Stokes polarization parameters, given by Bartschat et al. (1981), can be calculated using the spatial state multipoles $\langle T(1)^\dagger_{k_1q_1}\rangle$. Jitschin et al. (1984) found the values of $P_1\ P_2$ and $P_3$ as follows:

$$
P_1 = \frac{-(3)^{1/2}\langle T(1)^\dagger_{22}\rangle}{\langle T(1)^\dagger_{00}\rangle + \langle T(1)^\dagger_{20}\rangle} \tag{2.27a}
$$

$$
P_2 = 0 \tag{2.27b}
$$

$$
P_3 = (3/2)^{1/2}\frac{\langle T(1)^\dagger_{10}\rangle}{\langle T(1)^\dagger_{00}\rangle + \langle T(1)^\dagger_{20}\rangle} \tag{2.27c}
$$

By expanding the spatial state multipoles $\langle T(1)^{\dagger}_{k_1 q_1} \rangle$ in terms of (2.24) and retaining only the terms with the largest perturbation coefficient $\underline{G}$, taken from the Table 2.2, approximate values for (2.27a)–(2.27c) can be found as given below:

$$p_1 = (3)^{1/2} \frac{\langle T(1)^{\dagger}_{22} \rangle}{\langle T(1)^{\dagger}_{00} \rangle}$$

$$= -(3)^{1/2} \times 0.098 \times \frac{\langle T(L)^{\dagger}_{22} \times T(S)^{\dagger}_{00} \times T(I)^{\dagger}_{00} \rangle}{\langle T(L)^{\dagger}_{00} \times T(S)^{\dagger}_{00} \times T(I)^{\dagger}_{00} K \rangle} = P \qquad (2.28a)$$

$$P_2 = 0 \qquad (2.28b)$$

**Table 2.2** Perturbation coefficient $\underline{G}$ for sodium 3P state needed to calculate (2.25)

| $K_1$ | $q_1$ | $K_L$ | $Q_L$ | $K_S$ | $Q_S$ | $K_I$ | $Q_I$ | $\underline{G}\dfrac{Q_L\ Q_S\ Q_I\ q_1}{Q_L\ Q_S\ Q_I\ q_1}$ |
|---|---|---|---|---|---|---|---|---|
| 0 | 0 | 0 | 0 | 0 | 0 | 0 | 0 | 1.000 000 |
| 1 | 0 | 0 | 0 | 0 | 0 | 1 | 0 | 0.348,480 |
| 1 | 0 | 0 | 0 | 1 | 0 | 0 | 0 | 0.216,245 |
| 1 | 0 | 0 | 0 | 1 | 0 | 2 | 0 | −0.010,709 |
| 1 | 0 | 2 | 0 | 0 | 0 | 1 | 0 | −0.011,300 |
| 1 | 0 | 2 | 0 | 0 | 0 | 3 | 0 | −0.005,514 |
| 1 | 0 | 2 | 0 | 1 | 0 | 0 | 0 | −0.025,826 |
| 1 | 0 | 2 | 0 | 1 | 0 | 2 | 0 | −0.053,211 |
| 2 | 0 | 0 | 0 | 0 | 0 | 2 | 0 | 0.120,754 |
| 2 | 0 | 0 | 0 | 1 | 0 | 1 | 0 | 0.063,136 |
| 2 | 0 | 0 | 0 | 1 | 0 | 3 | 0 | 0.008,030 |
| 2 | 0 | 0 | 0 | 1 | ±1 | 1 | ±1 | 0.031,568 |
| 2 | 0 | 2 | 0 | 0 | 0 | 0 | 0 | 0.098,207 |
| 2 | 0 | 2 | 0 | 0 | 0 | 2 | 0 | 0.039,597 |
| 2 | 0 | 2 | 0 | 1 | 0 | 1 | 0 | 0.021,925 |
| 2 | 0 | 2 | 0 | 1 | 0 | 3 | 0 | −0.024,963 |
| 2 | 0 | 2 | 0 | 1 | ±1 | 1 | ±1 | −0.007,176 |
| 2 | 2 | 2 | 2 | 0 | 0 | 0 | 0 | 0.098,207 |
| 2 | 2 | 2 | 2 | 0 | 0 | 2 | 0 | −0.039,597 |
| 2 | 2 | 2 | 2 | 1 | 0 | 1 | 0 | 0.002,260 |
| 2 | 2 | 2 | 2 | 1 | 0 | 3 | 0 | 0.022,695 |
| 2 | 2 | 2 | 2 | 1 | ±1 | 1 | ±1 | −0.006,780 |

$$P_3 = (3/2)^{1/2} \left( \frac{\langle T(1)^\dagger_{10} \rangle}{\langle T(1)^\dagger_{00} \rangle} \right)$$

$$= (3/2)^{1/2} \times 0.348 \times \frac{\langle T(L)^\dagger_{00} \times T(S)^\dagger_{00} \times T(I)^\dagger_{00} \rangle}{\langle T(L)^\dagger_{00} \times T(S)^\dagger_{00} \times T(I)^\dagger_{00} \rangle} \qquad (2.28c)$$

$$+0.216 \times \frac{\langle T(L)^\dagger_{00} \times T(S)^\dagger_{00} \times T(I)^\dagger_{00} \rangle}{\langle T(L)^\dagger_{00} \times T(S)^\dagger_{00} \times T(I)^\dagger_{00} \rangle}$$

$$= (27/10)^{1/2} \times 0.348 P_1 + (3/2)^{1/2} \times 0.216 P'_s$$

The values of $P_1$ and $P_2$ from (2.28a) and (2.28b), respectively, are as expected from the theories given by Bartschat et al. (1981) and Bartschat and Blum (1982) while (2.28c) gives the circular polarization Stokes parameter $P_3$ which depends on two terms. The first term is due to the nuclear spin polarization $P_1$, which depends on the polarization of the atomic beam, whereas the second term is linked to the electron spin polarization $P_s$ of the collisionally excited state and thus bears information of the excitation process.

For a potassium atomic beam, polarized by using a hexapole magnet, the nuclear orientation alone yields a circular light polarization $P_3 = 0.120$. Taking into account the collisional interaction we can consider two limiting cases: first the spin-conserving direct and interference interaction would give $P_3 = 0.120 + 0.056 = 0.176$ and secondly the spin-reversing exchange interaction alone would yield $P_3 = 0.120 - 0.056 = 0.064$. The first case is generally realized at high collision energies while the second case can only be realized for differential measurements at certain angles (Moores and Norcross 1972).

The apparatus used, for these investigations, is described in Sects. 3.3.1–3.1.8, the experimental technique is outlined in the Sects. 4.5–4.10, the results and discussion are described in the Sects. 5.3.1–5.3.2 and the concluding remarks are given in the Chap. 6.

# Chapter 3
# Apparatus for the electron-atom collision studies

**Abstract** Research instruments such as photo detectors, charged particle energy analyzers and detectors and other apparatus used for the investigation of excitation and ionization of atoms and molecules by electron impact is described in detail. Also processes like inner shell ionization, X-rays, bremsstrahlung X-rays, negative ions, Augér and Coster–Kronig transitions which play important roles in these studies are discussed.

**Keywords** Analyzers · Detectors · Excitation · Ionization · Polarizers

## 3.1 Apparatus for the Electron-Atom Collision Studies

### 3.1.1 The Apparatus for Detecting Photons and Atomic Particles

Methods for the detection of atomic particles and photons have been developed to a high degree of sophistication: from intensity measurements with specific, geometric restrictions to angular coincidence and correlation measurements of atomic particles and photons including the analysis of spin effects. Latest achievements of such relevant measurements approached kinematically or even quantum mechanically "perfect" or "complete" scattering experiments from which collisional amplitudes and their phases can be derived (Kleinpoppen et al. 2005). Such highly informative experimental data represent most fundamental tests of basic collision theories in atomic physics.

We shall also refer to special schemes of applied apparatus mostly in connection with the relevant experimental data and results.

A. Chaudhry and H. Kleinpoppen, *Analysis of Excitation and Ionization of Atoms and Molecules by Electron Impact*, Springer Series on Atomic, Optical, and Plasma Physics 60, DOI 10.1007/978-1-4419-6947-7_3, © Springer Science+Business Media, LLC 2011

## 3.1.2   The Vacuum Chamber for Housing the Apparatus

Figure 3.1 shows a vacuum chamber and the apparatus inside. The vacuum chamber is a non-magnetic stainless steel cylinder, closed at one end, which has an outer diameter of 70 cm and a length of 60 cm. The chamber is mounted on a base flange, also made from non-magnetic stainless steel, and can be vacuum sealed with a viton "o" ring. A hoist is fixed to the closed end of the vacuum chamber which enables the lifting of the chamber and allowing access to the components of the apparatus inside the chamber.

A high vacuum of the order of $2 \times 10^{-7}$ torr can be maintained in the chamber with the help of a high-vacuum pump (e.g. Edwards MK2 diffstak), which uses Santovac 5 as a pumping fluid and has a pumping speed of 700 L/s for air.

**Fig. 3.1** The vacuum chamber

An Edwards rotary oil pump is used as the backing pump for the vacuum system. There is an oil trap between the backing pump and the diffusion pump to absorb the oil vapours from the backing line. In the event of a failure of cooling water or power, an interlock unit actuates an air-pressure-controlled butterfly valve which isolates the vacuum chamber from the pumping system. The diffusion pump is then automatically switched off and a magnetic valve isolates the rotary oil pump from the diffusion pump.

Figure 3.2 shows the apparatus inside the vacuum chamber. The electron gun (right), the Faraday cup (left), the ion analyzer (top) and the electron analyzer (at the back) are all supported on separate turntables to facilitate their independent placement.

**Fig. 3.2** The apparatus inside the vacuum chamber

### 3.1.3   The Electron Gun and the Faraday Cup

The electron gun used in these experiments is a directly heated tungsten filament type gun capable of producing a high-energy focused beam of electrons having a diameter of about 2 mm in the interaction region. Figure 3.3 illustrates the gun schematically. A hairpin-shaped tungsten filament is mounted on a ceramic support. The filament needs a heating current of about 2.7 A and can supply an electron beam current of several microamperes.

The electron gun assembly which is supported by ceramic rods is kept at high voltage while the voltage on the next electrode can be varied to focus the beam on the last (collimating) electrode which is kept at ground potential.

A Faraday cup collects the electron beam after its collisional interaction with the gas atoms. The Faraday cup consists of three cylinders of varying diameters and lengths and made from aluminium, which are supported on boron nitride spacers to keep them electrically isolated from one another. Figure 3.4 shows the Faraday cup assembly in detail. The Faraday cup has very low back-scattering and secondary electron emission and thus over 95% of the electrons reaching the Faraday cup are collected. Kuyatt (1968) has discussed various configurations of Faraday cups suitable for preventing secondary electron emission. To further minimize the secondary electron emission from different parts of the electron gun and Faraday

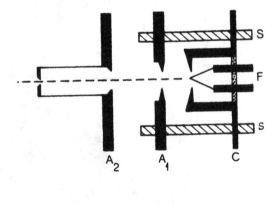

**Fig. 3.3** The electron gun. F is a hairpin-shaped filament, C the cathode, $A_1$ the focusing electrode, $A_2$ the collimating electrode and the parts marked S are the ceramic rods supporting the electrodes of the gun

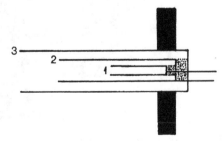

**Fig. 3.4** The Faraday cup. 1, 2 and 3 are the three concentrically placed aluminium cylinders separated by boron nitride spacers

cup, all these parts have been covered with soot. A Keithly electrometer measures the Faraday cup electron current.

### 3.1.4   The Atomic/Molecular Beam Source

A well-defined beam of atomic or molecular gases is obtained by effusing the gases through a nozzle consisting of a multi-capillary array (Lucas 1973). Each capillary is about 3 mm long and has an internal diameter of 0.05 mm. The multi-capillary array is encased in a long demountable aluminium capsule, which is welded to a brass tube having an inner diameter of about 6 mm. This brass tube is carrying the gas nozzle, which enters the vacuum chamber through the base flange along the axis of rotation of the turntables carrying the electron gun, the Faraday cup and the electron analyzer. Gas, at low pressure, is supplied to this assembly from a pressure regulating system consisting of a needle valve and a pressure regulator fixed to the gas cylinder. Any gas pressure can be obtained inside the vacuum chamber by regulating the gas supply system.

Gas pressure in the collision region is adjusted to be low enough to ensure single collision conditions (Nagy et al. 1980). For helium, neon and argon, the pressure in the interaction region is kept at about $5.0 \times 10^{-6}$ torr while for krypton and xenon the pressure is about $4.0 \times 10^{-6}$ torr. Research grade gases supplied by British Oxygen Limited are used.

### 3.1.5   The Electron Analyzer

Figure 3.5 shows schematically the 30° parallel plate electrostatic analyzer which has been used for the analysis of electrons ejected as a result of the collisional interaction of electrons with the gaseous atoms. The analyzer has been built from

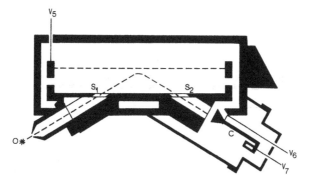

**Fig. 3.5**  The electron analyzer. O is the interaction region, $S_1$ and $S_2$ are the defining slits and C is the channeltron used for the detection of electrons

non-magnetic stainless steel. A wire mesh, made from 0.1 mm thick tungsten wire, spot-welded 1 mm apart on a supporting frame, acts on the upper plate for the deflection of electrons inside the analyzer. The potential difference between the mesh and the lower plate defines the energy of the selected electrons. The analyzer elements and the channeltron are covered by a light-tight shielding to avoid stray electrons and photons entering the analyzer. Boron nitride spacers are used for the insulation of the upper plate and the channeltron.

The trajectory of electrons of energy $E = eU$, and making an angle $\theta$ with the $x$-axis, is shown in Fig. 3.6. If $d$ is the separation of the plates, $V$ the potential difference between the plates I and II, $h$ and $y$ are the distances of the source and the image from plate I, then the equation describing the electron path in the field-free path inside the analyzer is (Green and Proca 1970)

$$X(\theta, V, h) = (h + y) \cot \theta + \left(\frac{2dU}{V}\right) \sin 2\theta, \qquad (3.1)$$

where $X$ is measured from the end of the analyzer as shown in Fig. 3.6. For the analyzer to have first- and second-order focusing properties, it is required that

$$\frac{\partial x}{\partial \theta} = 0, \quad \frac{\partial^2 x}{\partial^2 \theta^2} = 0, \quad \theta = 30° \quad \text{and} \quad \frac{dU}{hV} = 2.$$

The ratio $U/V$ is called the analyzer factor, $f$. Thus,

$$f = \frac{U}{V} = \frac{2h}{d}. \qquad (3.2)$$

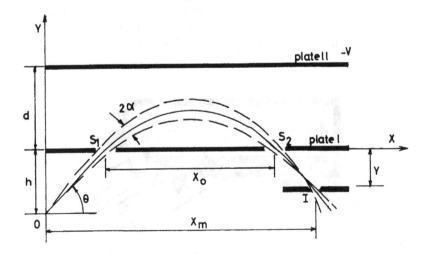

**Fig. 3.6** Electron path in the parallel plate electrostatic analyzer shown in Fig. 3.5

According to Schmitz and Mehlhorn (1972), any parallel plate analyzer will have second-order focusing property if it satisfies the condition $0 < y < h$.

The analyzer resolution is given by Green and Proca (1970)

$$\frac{\Delta U}{U} \cong 1.6(\alpha)^3 \tag{3.3}$$

for a 30° parallel plate analyzer, where $\alpha$ is the angle as shown in Fig. 3.6. Green and Proca (1970) have compared this analyzer with other analyzers and have shown that while this analyzer has double focusing property, other types of analyzers are mostly monochromators.

The analyzer transmission, $T$, is given by

$$T = \frac{\Delta\varphi}{4\pi}(\sin\theta_{max} - \sin\theta_{min}) = \left(\frac{\delta\varphi}{2\pi}\right)\cos\bar{\theta}\sin\alpha, \tag{3.4}$$

where $\Delta\phi = 2\pi$, the transmission is given by

$$T = \frac{\sqrt{3}}{2}\sin\alpha. \tag{3.5}$$

Green and Proca (1970) and Proca and Green (1970) have given a detailed discussion about other characteristics of this analyzer.

As shown in Fig. 3.5, the negative voltage $V_5$ given to the grid of the analyzer determines the energy of the electrons which are focused at the cone of the channeltron. The lower plate of the analyzer is kept at the ground potential. As the detection efficiency of a channeltron varies with the energy of the detected electrons, for maximum detection efficiency, it is necessary to ensure that the electrons are incident at the surface of the cone of the channeltron with a constant energy of 200 eV. To achieve this condition, $V_6$ is made positive or negative and is adjusted to keep the incident electron energy equal to 200 eV. In addition, the voltage $V_7$ is adjusted to keep the potential difference between $V_6$ and $V_7$ the same throughout the experiment, thus keeping the gain of the channeltron nearly constant.

## 3.1.6   The Ion Analyzer

The ions formed as a result of ionization are analyzed with respect to their charge (mass being the same) by a time-of-flight (TOF) method. Ions from the interaction region are pulled into the ion analyzer by a weak electric field. These ions are accelerated further until they acquire the desired kinetic energy, whereupon they enter the analyzer and are allowed to drift in a field-free region ($\sim$35 mm). Finally, they are accelerated to several kiloelectron volt energy before they impinge on the channeltron surface for detection.

Figure 3.7 shows schematically the ion analyzer. The analyzer, essentially, consists of two concentrically placed cylindrical tubes, made out of non-magnetic stainless steel, having grids at their ends formed by welding together 0.2 mm diameter tungsten wire. These tubes are fixed to aluminium housing for the channeltron (Mullard 913B). Spacers are used to insulate the drift tubes from the channeltron housing and from each other.

If the distance $D$ travelled by ions inside the drift tube is large compared with the distance between the interaction region and the tube opening, then for an ion of mass $M$, charge $e$ and velocity $v_n$, the TOF $t_n$ is given by

$$t_n = \frac{D}{v_n} = \left(\frac{MD^2}{2eV}\right)^{1/2} \times \frac{1}{n^{1/2}}, \qquad (3.6)$$

where $V$ is the voltage through which the ion has been accelerated to attain velocity $v_n$.

For an experiment with any gas, the quantity $(MD^2/2eV)$ remains constant and the TOF $t_n$ is thus indirectly proportional to $n^{1/2}$. Thus, with mass being the same, ions having different charge states should separate out in the form of a spectrum, such as the one shown in Fig. 3.8.

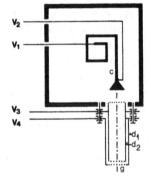

**Fig. 3.7** The ion analyzer. $d_1$ and $d_2$ are the two concentric cylindrical drift tubes, insulated from each other. The tungsten wire grids are marked as g. c is a channeltron for the detection of ions

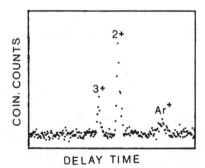

**Fig. 3.8** A typical time-of-flight (TOF) spectrum of argon ions. Peaks for $Ar^+$, $Ar^{2+}$ and $Ar^{3+}$ ions can be seen in the spectrum

Any initial kinetic energy that an ion possesses, whether thermal energy of the parent atom or kinetic energy released in the formation of an ion, has a strong effect on the flight time. An ion with initial velocity in the forward direction will arrive too soon while one which starts off in the opposite direction must be retarded to stop, then returned to its starting position with a reversed velocity. The difference in arrival times between these two extreme ions is just twice the retardation time. The effect of initial kinetic energy in TOF mass spectrometers has been studied by Franklin et al. (1967) who have shown that for small initial kinetic energies compared with the energy acquired in the field, the greatest and the least flight times are symmetrical about the flight time of an ion from the centre of the interaction region and with zero initial velocity. There is then one-to-one correspondence between arrival times and initial velocities in the flight direction and, for a Maxwellian initial energy distribution, the shape of the peak in a TOF spectrum is a Gaussian with a width at half maximum ($W_{1/2}$) given by

$$W_{1/2} = \frac{1.664(2kTM)^{1/2}}{eE},$$ (3.7)

where $k$ is the Boltzman constant, $T$ is the temperature (K), $m$ is the mass of ion in atomic mass units and $E$ is the electric field in volt per centimetre.

This result is applicable to the TOF coincidence spectroscopy. Equation (3.7) shows that a thermal ion of mass 40 a.u. at 300 K in a field of 15 V/cm should have a delay half-width of about 0.2 μs. This is a fundamental limitation on the coincidence resolving time. The spread of ion flight time can be reduced by increasing the extraction field in the interaction region, but doing so necessarily worsens the electron energy resolution. The ionization region itself must have a definite width, $\delta S$, in the field direction which gives electrons a maximum energy spread, $E \times \delta S$. The conditions of any experiment must be a compromise, chosen between time resolution for coincidence and energy resolution for electrons. The situation can be improved by making the ionization region as small as possible and by providing the target gas in the form of an atomic beam.

A low voltage (between 15 and 25 V) draws the produced ions out of the interaction region while another voltage (between 25 and 100 V) gives the ions a high enough drift energy. At the end of the ion drift region, a suitable high voltage gives the ions enough energy to be efficiently detected in the channeltron. Ravon (1982) has shown how the detection efficiency of a channeltron varies for various ions having different charge states and different energies and that for almost all ions, having energy of 4 keV, the detection efficiency exceeds 80%.

## 3.1.7 The Negative Ions

The negative ions of atoms can be specified in analogy to positive ions. In general, however, the binding force is only strong enough to bind one electron to the neutral atom. The binding energy of the additional electron bound to the neutral atom (A) is called the electron affinity $E_{EA}(A)$ and is defined quantitatively as follows:

$$E_{EA}(A) = E_{tot}(A) - E_{tot}(A^-)$$

with $E_{tot}(A)$ and $E_{tot}(A^-)$ are the total energies of the neutral atom and the negative ion, respectively. The electron affinity is positive if the negative ion is stable. Table 3.1 summarizes electron affinity for atoms with nuclear charges $Z \leq 86$. Atoms with closed $ns^2$ and $np^6$ sub-shells have no stable negative ions in their ground states.

Obviously, the hydrogen atom can have a stable negative ion with the configuration $H^-$ ($1s^2$) since the Pauli's exclusion principle would not forbid it; the two electrons would have anti-parallel spin directions as, for example, in the ground state of helium. Indeed the negative hydrogen ion plays an important role in various aspects of atomic physics and in the atmospheres of stars (e.g. in absorption of electromagnetic radiation by $H^-$ in the photospheres of the sun and other cooler stars). According to the Pauli's principle, the helium atom cannot bind another electron in the $n = 1$ state, accordingly it does not have a stable negative ion. This argument is also valid for all other rare gas atoms with closed $np^6$ sub-shells and all earth alkaline metal atoms and for zinc, cadmium and mercury with closed electron configuration $ns^2$. By satisfying the Pauli's principle, an additional electron can only be bound in higher excited quantum states of atoms with closed shells in the ground state. Such negative ions in excited quantum states play important roles in atomic collision processes, particularly in electron-atom scattering in which excited negative ions are produced at sharp resonant energies of the incoming electrons. These excited negative ions can be metastable with life times of the order of microseconds, e.g. for the negative ion state $He^-$ ($1s\ 2s\ 2p^4\ P_0$) or very short life times, e.g. $10^{-13}$ to $10^{-14}$ s for $H^-$ ($1snl$). The decay of such excited negative ions is caused by the process of autoionization, i.e. the negative ion decays into a neutral atom and an electron. Such processes take place most quickly under the action of the Coulomb interaction as for the states with very short lifetimes, but proceed very slowly under the action of spin–spin or spin–orbit interaction or long-lived metastable states.

Relatively large electron affinities of the halogens, F, Cl, Br and I, are particularly noticeable. This indicates that atoms in the configuration $np^5$ have a strong tendency (i.e. affinity) to fill their outer shells completely. Negative ions play important roles in electrical discharges and absorption processes of planetary and star atmospheres. Negative ions are also used in particle accelerators.

After acceleration to a high potential by a static electric field, negative ions pass through a thin polymer foil, in which their electrons are stripped off converting them into positive ions. Subsequently, these ions are accelerated by an electrode at earth or negative potential.

**Table 3.1** Electron affinities (in eV) of atoms (<0 means that a negative ion is not stable): (a) the principal groups and (b) the long rows of the periodic system

**a**

| $^{1}$H 0.754209 | | | | | | | $^{2}$He <0 |
|---|---|---|---|---|---|---|---|
| $^{3}$Li 0.6180 | $^{4}$Be <0 | $^{5}$B 0.277 | $^{6}$C 1.2629 | $^{7}$N <0 | $^{8}$O 1.4611215 | $^{9}$F 3.399 | $^{10}$Ne <0 |
| $^{11}$Na 0.54793 | $^{12}$Mg <0 | $^{13}$Al 0.441 | $^{14}$Si 1.385 | $^{15}$P 0.7465 | $^{16}$S 2.077120 | $^{17}$Cl 3.617 | $^{18}$Ar <0 |
| $^{19}$K 0.50147 | $^{20}$Ca <0 | $^{31}$Ga 0.3 | $^{32}$Ge 1.2 | $^{33}$As 0.81 | $^{34}$Se 2.02069 | $^{35}$Br 3.365 | $^{36}$Kr <0 |
| $^{37}$Rb 0.48592 | $^{38}$Sr <0 | $^{49}$In 0.3 | $^{50}$Sn 1.2 | $^{51}$Sb 1.07 | $^{52}$Te 1.9708 | $^{53}$I 3.0591 | $^{54}$Xe <0 |
| $^{55}$Cs 0.47163 | $^{56}$Ba <0 | $^{81}$Tl 0.2 | $^{82}$Pb 0.364 | $^{83}$Bi 0.946 | $^{84}$Po 1.9 | $^{85}$At 2.8 | $^{86}$Rn <0 |

**b**

| $^{20}$Ca | $^{21}$Sc | $^{22}$Ti | $^{23}$V | $^{24}$Cr | $^{25}$Mn | $^{26}$Fe | $^{27}$Co | $^{28}$Ni | $^{29}$Cu | $^{30}$Zn |
|---|---|---|---|---|---|---|---|---|---|---|
| <0 | 0.188 | 0.079 | 0.525 | 0.666 | <0 | 0.163 | 0.661 | 1.156 | 1.228 | <0 |
| $^{38}$Sr | $^{39}$Y | $^{40}$Zr | $^{41}$Nb | $^{42}$Mo | $^{43}$Tc | $^{44}$Ru | $^{45}$Rh | $^{46}$Pd | $^{47}$Ag | $^{48}$Cd |
| <0 | 0.307 | 0.426 | 0.893 | 0.746 | 0.55 | 1.05 | 1.137 | 0.557 | 1.302 | <0 |
| $^{56}$Ba | $^{57}$La | $^{72}$Hf | $^{73}$Ta | $^{74}$W | $^{75}$Re | $^{76}$Os | $^{77}$Ir | $^{78}$Pt | $^{79}$Au | $^{80}$Hg |
| <0 | 0.5 | $\approx 0$ | 0.322 | 0.815 | 0.15 | 1.1 | 1.565 | 2.128 | 2.30863 | <0 |

## 3.1.8 High-Voltage Power Supplies, Multi-channel Analyzer and Other Electronic Equipment

For the purpose of accelerating electrons, a high voltage with a negative polarity is supplied to the cathode of the electron gun by a regulated (0.2%) Universal Voltronics (Model BRE-20.25-R2) power supply. The filament heating power supply and the anode power supply are isolated using an isolation transformer. An E.G.& G. Ortec multi-channel analyzer (MCA) is used in the pulse-height analysis (PHA) mode to record the time spectra for the coincidence studies. Standard NIM units (supplied by Ortec and Nuclear Enterprise) are used for signal amplification and discrimination, etc.

## 3.1.9 Hyper Pure Germanium (HPGe) X-Ray Detector

The X-ray detector used for spectroscopy is an E.G.& G. Ortec 1000 series low-energy photon spectrometer (LEPS). It is essentially a diode made from hyper pure germanium (HPGe) crystal (impurities less than $10^{10}$ atoms/cm$^3$) and is operated at liquid nitrogen temperature. The detector has the dimensions of 6 mm active diameter by 5 mm active depth. The Beryllium entrance window is 0.0254 mm thick. The front electrode of the detector is ion implanted to obtain a thin, robust and reliable contact.

When an X-ray photon is absorbed in the detector, it produces electron hole pairs (on average it require 2.95 eV per pair) which are swept out of the detector volume by the electric field due to the bias voltage (1,200 V) and the resulting current pulse is integrated by a charge-sensitive preamplifier. The first transistor of the preamplifier is cooled to minimize electronic noise. The amplitude of the integrated current pulse is proportional to the energy lost by the incident X-ray inside the detector. Figure 3.9 shows a graph of detection efficiency versus photon energy for the HPGe detectors of various sizes having windows of different thickness.

### 3.1.10   Energy Structure of Inner Shells, X-Ray Spectra, Augér Effect and Coster-Kronig Transitions

Optical radiation transitions take place by energy changes of electrons in the outer most shells. The quantum numbers of the electron involved in such transitions of absorption or emission are changed. Transitions of such types within or between closed shells or sub-shells of the electrons of the atom are not possible since the total electron configuration of the shell or sub-shell would not be expected to change. As a consequence, the total energy of closed shells or sub-shells must remain constant. Accordingly, the energy of the closed shells or sub-shells of the electrons cannot be measured directly.

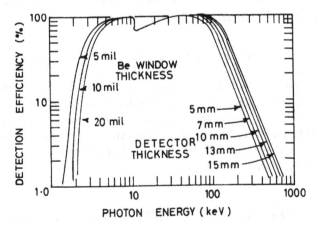

**Fig. 3.9** Detection efficiency curves for HPGe detectors of different thickness and having beryllium windows of different thickness (1 mil = 0.025 mm)

However, the discovery of characteristic and continuous X-rays (Röntgen 1896) has developed into an analysis of the energy structure of inner shells of atoms. Characteristic X-rays are monoenergetic photons in the wavelength region from about 0.01 to 10 nm. By means of special mechanisms, which are based on photo-absorption or collisions between atoms and electrons, ions or other particles, electrons can be knocked out of their inner shells. Consequently, the atom is ionized in an inner shell. In other words, an electron of a previously full and closed sub-shell is now missing. For example, an electron of the $1s^2$ sub-shell may have been knocked out in such a way that the resulting ion has the configuration $A^+$ ($1s \, 2s^2 \, 2p^6 \ldots$); one describes this ionic state by saying that the K-shell of atom has an electron hole.

Electron holes can also be produced in L, M, N... shells. The K-shell electrons are most strongly bound by the atomic nucleus; electrons removed from the L, M, N... shells are bound with decreasing strength. Electron holes in the shells can be refilled by a mechanism in which an electron from a higher shell *jumps* into a lower shell; for example, a gap in the K-shell can be filled by an electron of the L-shell or the M, N, . . ., shell. Since an electron in the L-shell is more weakly bound than an electron of the K-shell, the excess energy of the ion is released in the form of a photon; we can describe this process by the following reaction:

$$A^+(K) \rightarrow A^+(L) + h\upsilon_{KL}.$$

The ion $A^+(K)$ with its electron gap in the K-shell is transferred into an ion $A^+(L)$ with a gap in the L-shell. In other words, the initial electron hole in the K-shell is filled by an electron of the L-shell whereby a characteristic X-ray line radiation with the energy $h\upsilon_{KL}$ is produced. The electron hole of the K-shell is passed onto a gap of the L-shell. The electron hole in the L-shell can successively be filled up by an electron of the M, N,... shells with the emission of characteristic X-ray lines $h\upsilon_{LM}$, $h\upsilon_{LN}$, . . ..

The interpretation of the characteristic X-ray line is surprisingly simple; we shall deal theoretically with the energy levels associated with the characteristic lines in a hydrogen-like approximation. We introduce an effective nuclear charge $Z_{eff} = Z - \sigma_n$ with a screening constant $\sigma_n$ for a given value of Z. The energy of removing an electron from an inner shell of principal quantum number $n$ and effective nuclear charge $Z_{eff}$ is then approximately described by a modified Bohr formula

$$E_n = -Rhe \frac{(Z - \sigma_n)^2}{n^2}, \tag{3.8}$$

where $n = 1, 2, 3, 4, \ldots$ is associated with the K, L, M, N shells . . . ., respectively. This formula describes the gross structure of the energy needed to remove an electron from an inner shell; a less precise way of expressing this is to speak of the "energy $E_n$ of the shell."

Again in analogy to the fine structure of valence electrons of atoms, a fine structure for the energy of inner shells can be formulated. We describe the fine structure of electrons of inner shells in a hydrogen-like manner, i.e. we apply either the Sommerfeld or the quantum-mechanical fine-structure formula with a screening constant $s_n$ and an effective nuclear charge $Z_{eff} = Z - s_n$; we then obtain

$$E_{j,l} = -\frac{hcR\alpha^2(Z - s_n)^4}{n^3}\left[\frac{1}{j + (1/2)} - \frac{3}{4\pi}\right] \tag{3.9}$$

the quantum numbers $j$, $l$ and $s$ with $j = l + 1/2$ are the same as for atomic hydrogen. The various inner shells have a hydrogen-like gross and fine structure for the respective energies of their electron holes. A missing electron in the K-shell is described by a $1s^2\ S_{1/2}$ configuration. There are three possibilities for an electron hole in the L-shell, in the configuration $2s\ 2p^6\ ^2S_{1/2}$, $2s^2\ 2p^5\ ^2P_{1/2}$ and $2s^2\ 2p^5\ ^2P_{3/2}$. The 2s electron hole has the orbital quantum number $l = 0$ while the $2p^5$ electron hole has $l = 1$. Accordingly, the level symbols $^2S_{1/2}$ and $^2P_{1/2,3/2}$ are appropriate and justified for the description of the resulting configuration of the electron holes. The spin–orbit interaction in the electron holes is described by the usual notation for the structure splitting.

We continue to apply this scheme of level symbols for the M-shell; in the M-shell, we have the configuration $3s\ 3p^6\ 3d^{10}\ ^2S_{1/2}$, $3s^2\ 3p^5\ 3d^{10}\ ^2P_{1/2,3/2}$ and $3s^2$ $3p^6\ 3d^9\ ^2D_{3/2,5/2}$. In the N-shell, we have the additional $^2F_{5/2,7/2}$ configurations and in the O-shell the $^2G_{7/2,9/2}$ configurations.

Another consequence of the hydrogen-like description of inner shell states is the fact that transition rules for the electron holes are identical to the optical selection rules for one-electron atoms, i.e. $\Delta l = \pm 1$, $\Delta j = 0, \pm 1$ and $\Delta n$ is arbitrary. Traditionally, special notations for the energies of shells and sub-shells and their transitions have been in existence for a long time; these are summarized in Table 3.2. According to (3.8) and (3.9), the K-shell has only one energy state while the L-shell splits into three sub-shells $L_I$, $L_{II}$ and $L_{III}$, which are described by the electron configurations $2^2S_{1/2}$, $2^2P_{1/2}$ and $2^2P_{3/2}$. Similarly, the M-shell has $M_I$, ......., $M_V$ sub-shells with the corresponding electron configurations as described above which are also listed in the table. According to the selection rules, transitions between K-shell and the L-shell are allowed only between the K-shell and LII and LIII sub-shells; the X-ray lines of these transitions are called $K_{\alpha_1}$ and $K_{\alpha_2}$. Further, K and L lines are given in Table 3.2.

So far, we have given the systematics of the energy structure of inner shells and their possible transitions. We now consider experimental results. The energy structure of the inner electron shells can be studied by absorption measurements of short wave electromagnetic radiation or by observing the emission of X-rays produced in impact excitation. The arrangement of a classical X-ray tube is illustrated in Fig. 3.10. X-rays are produced in this tube by energetic electrons striking the anode. Since the inner shells of atoms are to a large extent independent of solid-state interaction, it is expected that the origin of emission of characteristic

**Table 3.2** Nomenclature of K-, L- and M-shells with the total configurations $1\,^2S_{1/2}$, $2\,^2S_{1/2}$, $2\,^2P_{1/2}$ …. and the notations of the associated X-ray lines at the crossing point of the coordinate axis

| Electron configuration | Shells | K lines | L lines | | |
|---|---|---|---|---|---|
| | | K | $L_I$ | $L_{II}$ | $L_{III}$ |
| $1\,^2S_{1/2}$ | K | | | | |
| $2\,^2S_{1/2}$ | $L_I$ | | | | |
| $2\,^2P_{1/2}$ | $L_{II}$ | $\alpha_2$ | | | |
| $2\,^2P_{3/2}$ | $L_{III}$ | $\alpha_1$ | | | |
| $3\,^2S_{1/2}$ | $M_I$ | | | $\eta$ | $l$ |
| $3\,^2P_{1/2}$ | $M_{II}$ | $\beta_1$ | $\beta_4$ | | |
| $3\,^2P_{3/2}$ | $M_{III}$ | $\beta_1$ | $\beta_3$ | | |
| $3\,^2D_{3/2}$ | $M_{Iv}$ | | $\beta_{10}$ | $\beta_1$ | $\beta_2$ |
| $3\,^2D_{5/2}$ | $M_v$ | | $\beta_9$ | | $\alpha_1$ |
| $4\,^2S_{1/2}$ | $N_1$ | | | | |
| $4\,^2P_{1/2}$ | $N_{II}$ | $\beta_2$ | $\gamma_2$ | | |
| $4\,^2P_{3/2}$ | $N_{III}$ | $\beta_2$ | $\gamma_3$ | | |
| $4\,^2D_{3/2}$ | $N_{Iv}$ | | | $\gamma_1$ | $\beta_{15}$ |
| $4\,^2D_{5/2}$ | $N_v$ | | | | $\beta_2$ |
| $4\,^2F_{5/2}$ | $N_{VI}$ | | | | |
| $4\,^2F_{7/2}$ | $N_{Vii}$ | | | | |
| $5\,^2S_{1/2}$ | $O_1$ | | | $\gamma_8$ | $\beta_7$ |
| $5\,^2P_{1/2}$ | $O_{II}$ | | $\left.\vphantom{\begin{array}{c}a\\b\end{array}}\right\}\gamma_4$ | | |
| $5\,^2P_{3/2}$ | $O_{III}$ | | | | |
| $5\,^2D_{3/2}$ | $O_{Iv}$ | | | $\gamma_6$ | $\left.\vphantom{\begin{array}{c}a\\b\end{array}}\right\}\beta_5$ |
| $5\,^2D_{5/2}$ | $O_v$ | | | | |

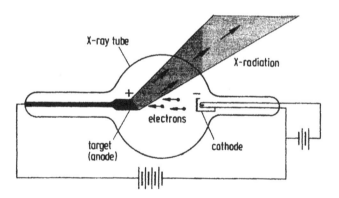

**Fig. 3.10** Schematic arrangement of an X-ray tube with the cathode at a negative potential and the anode at a positive potential. X-radiation is produced when electrons strike the anode

X-rays in such an x-tube is approximately that of free atoms. However, we now draw attention to modern methods of producing X-radiation by impact excitation of free atoms (completely free of a solid-state environment). Figure 3.11 shows typical examples of *X-ray spectra*. While the spectra of chromium and tungsten are continuous in their shape with a maximum but with the other structure, the molybdenum spectrum has two sharp peaks. These peaks are the $K_\beta$ line at 0.06 nm and the $K_\alpha$ line at 0.07 nm; they are not resolved into their components, i.e. in

**Fig. 3.11** X-ray emission
spectra of chromium (Cr),
molybdenum (Mo) and
tungsten (W) produced by 35
keV-electrons. The (cut-off)
peaks at 0.06 and 0.07 nm
represent the $K_\alpha$ and $K_\beta$ lines
of molybdenum. The intensity
$I$ as a function of the
wavelength $\lambda$ is given in
arbitrary units $I_1$ (after Urey
1918)

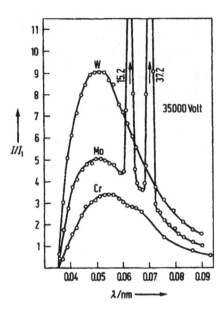

$K_{\alpha 1}$ and $K_{\alpha 2}$ or $K_{\beta 1}$ and $K_{\beta 2}$ according to Table 3.2. The continuous spectra of chromium and tungsten and the continuous parts of molybdenum spectrum are called *Bremsstrahlung spectra of X-radiation* (German for "deceleration radiation") of these elements produced by the electrons in the material of the anode of the X-ray tube. In general, the contributions from the Bremsstrahlung and the characteristic X-radiation superimpose on each other in X-ray spectra. In Fig. 3.11, we observe a common threshold value at the short wavelength end of the three Bremsstrahlung spectra. The explanation for this is that at this point the photon energy equals the kinetic energy ($eU$) of the electrons striking the anode. Taking account of the energy conservation law, the maximum possible energy of the Bremsstrahlung photons should obey the relation $h\nu_{max} = eU$.

This energy relation for the limiting frequency $\nu_{max}$ has been fully confirmed.

Apart from the emission spectroscopy of characteristic X-radiation, absorption spectroscopy has also been applied. Similarly as for light in the optical range of atomic spectroscopy, X-rays can also experience a reduction of initial intensity $I_o$ when they pass through the target. The typical exponential law for the measured intensity behind a target of thickness $x$ is $I = I_o e^{-\mu x}$ with the absorption coefficient $\mu$ of the X-radiation which decreases monotonically with increasing frequency or energy (see Fig. 3.12). Therefore, X-rays of high energy (so-called "hard" X-rays) have a greater chance of passing through matter than those of lower energy ("soft" X-rays). However, if the energy of the incident X-radiation is high enough, electrons from K or L, M, . . . .., shells may be knocked out. The absorption cross-section $Q = \mu/n$ ($n$ is the atomic density) shows an abrupt increase at the energies of these shells; they are called the absorption edges of the K, L, M,. . ..., shells. With X-ray frequencies $\nu_K, \nu_L, \ldots\ldots$ and energies $E_n$ of the shells, i.e. $h\nu_K, h\nu_L, \ldots\ldots$, we

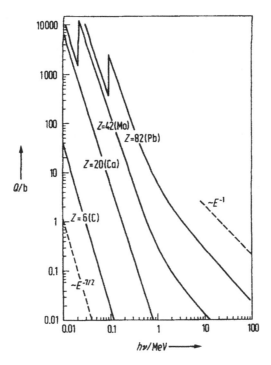

**Fig. 3.12** Cross-section $Q = \mu/n$ ($\mu$ is the absorption coefficient and $n$ the density of atoms) of photo-ionization as a function of the photon energy $h\nu$ for heavy, medium-heavy and light atoms. The slopes of the functions of these cross-sections are indicated by $E^{-1/2}$ and $E^{-1}$ in the figure, respectively, they refer to the Born approximation for which the photon energy $h\nu$ is expected to be large compared with the binding energy of the knocked-out electron, i.e. $h\nu \gg E_{J,n}$, the other case requires that the photon energy is large compared with the rest energy of the electron, i.e. $h\nu \gg mc^2$. The unit b = barn = $10^{-24}$ cm$^2$ is usually applied in nuclear and particle physics. The absorption edges, the sharp peaks at 20 keV for Mo and about 80 keV for Pb, indicate the onset of the K-shell absorption

define (as in optical spectroscopy) the term values $T_n$ for the states of inner shells by the relationship $T_n = -E_n/hc = R((Z - \sigma_n)^2)/n^2$ [see (3.8)]. The wave numbers for allowed transitions between the shells follow from the combination principle; i.e. $\bar{\nu} = 1/\lambda = T_n - T_{n'}$. In other words, in emission spectroscopy, the wave numbers determine the term differences whereas the term values $T_n$ or energies $E_n$ are measured in absorption spectroscopy. This description is further simplified by the experimental observation that the wave numbers of the K, L, ....., spectral series for the characteristic lines follow the Moseley law. Instead of describing the term differences of the K, L, ..., spectral lines of characteristic X-radiation by the formula

$$\bar{\nu} = \frac{1}{\lambda} = R\left[ \frac{(Z - \sigma_{n'})^2}{n'^2} - \frac{(Z - \sigma_{n''})^2}{n''^2} \right], \tag{3.10}$$

only one value of the screening constant $\sigma$ is required, to a good approximation; for example, the equation $\bar{v}_K = 1/\lambda = R(Z - \sigma_{n=1})(1 - 1/n^2)$ for the K-series gives $\bar{v}_{K_\alpha} = (3/4)R(Z - \sigma_{n=1})^2$ for the $K_\alpha$-lines. By applying Moseley's law, we plot the quantity $\sqrt{\bar{v}/R}$ as a function of the atomic number $Z$ (see Figs. 3.13 and 3.14). The screening constant is approximately $\sigma_{n=1} \approx 1$ for the K-series and $\sigma_{n=2} \approx 5.8$ for the L-series, however, these values do vary to a certain degree over the range of $Z$. The accuracy of the wave number measurements of characteristic X-ray lines is high enough to separate fine-structure effects in the Moseley diagrams. It has been found that it is possible to describe the fine-structure levels of inner shells by their "effective" atomic numbers $Z - \sigma_n$ and $Z - s_n$, i.e. with their screening constants $\sigma_n$ and $s_n$, respectively:

$$T_{n,l,j} = R(Z - \sigma_n)^2 - \frac{R\alpha^2(Z - s_n)^4}{n^3}\left[\frac{3}{4\pi} - \frac{1}{j + (1/2)}\right] \tag{3.11}$$

The second term of this equation is identical to Sommerfeld's formula for the fine structure with a screening constant $s_n$.

We have already discussed that $\sigma_n$ depends on $j$. The second term in (3.11) with the doublet splitting $j = l \pm (1/2)$ depends on the fourth power of $(Z - s_n)$ and describes the spin or regular doublets which correspond to the normal fine structure of optical spectra, e.g. $^2P_{1/2,3/2}$, $^2D_{3/2,5/2},\ldots$. Two levels with the same $n$ and $j$ but different $l$ (i.e. $l + 1/2$ or $l - 1/2$) experience a further splitting into irregular or screening doublets. This splitting is due to the observation that $\sigma$ and $s$ depend on the quantum number $l$, where the first term of (3.11) dominates over the second one.

Instead of describing the characteristic X-ray lines by the values $\sqrt{\bar{v}/R}$ in a Moseley diagram, one can use the value $\sqrt{T/R}$ in a Bohr-Coster diagram (see Fig. 3.15) which can be obtained directly from measurements of the K-, L- and

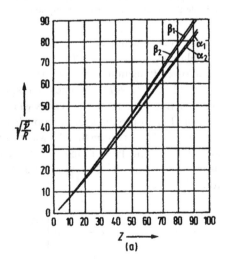

**Fig. 3.13** Moseley's law for
X-ray lines of the K-series

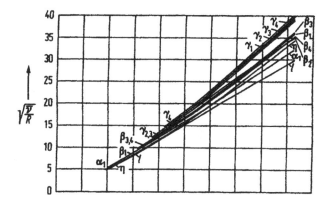

**Fig. 3.14** Moseley's law for X-ray lines of the L-series

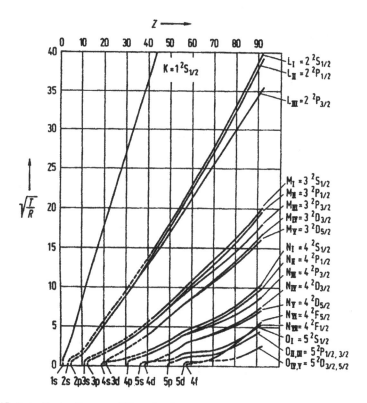

**Fig. 3.15** Bohr-Coster diagram of X-ray levels

M-absorption edges (see Fig. 3.12). By differentiating (3.11) with respect to $\sigma_n$ we obtain $\Delta T = ((2R(Z - \sigma))/n^2)\Delta\sigma$, which shows that the term differences for screening doublets is a linear function of the atomic number $Z$. Two terms with equal $j$ and $n$ but different $l$ run parallel to each other in a Bohr-Coster diagram as shown, for example, in Fig. 3.15 for the $L_I$, $L_{II}$ and $M_{II}$ ....states. The approximate linear dependence of screening doublets on $Z$ has some similarity to the optical spectra of iso-electronic sequences of ions. We also like to draw attention to the observation that the screening constants ($\sigma_n$) for the spectral series of characteristic X-ray lines and the absorption edges ($\sigma_K$, $\sigma_L$,...) can be different from each other. While $\sigma_{n=1} \approx 1$ is valid for the K-spectral series of characteristic X-ray lines, $\sigma_K \approx 2$–3 for absorption edges. This is understandable, since screening of characteristic K-shell transitions is only effective in the region near the atomic nucleus, on the other hand in a measurement of the absorption edge, an average screening from the shell on the edge of the ionization limit has to be taken into account.

As an example, Fig. 3.16 shows the X-ray energies of cadmium with allowed transitions, we note that in this representation, positive values for the wave numbers start at zero point coinciding with the ground state of the neutral atom.

For intensity reasons, solid-state targets have been used in most previous investigations. In recent experiments, however, it is possible to use free atoms as targets. The structure of such absorption edges is illustrated for the case of free rare gas atoms in Fig. 3.17; it consists of two components, firstly, of the contribution of photo-ionization, in which the ion produced is in the ground state and secondly of the contribution in which the atom is not ionized but excited into an unpopulated optical state of the atom. If the lifetime of the electron hole is sufficiently large and, accordingly, its energy width is small enough, a second edge may appear as, for example, for the Ar edge in Fig. 3.17 which is composed of a superposition of the two contributions discussed.

Apart from the characteristic X-ray lines discussed so far, so-called satellite and hyper-satellite lines are very small in comparison with the "normal" X-ray line intensities. Some of the satellite lines could be identified as resulting from quadrupole transitions; the share of such electromagnetic transitions increases, in comparison to dipole transitions, as the square of the wave number. However, the physical origin for the majority of the satellite and hyper-satellite lines is due to the production of double-electron holes (for satellite lines) or even several electron holes (for hyper-satellite lines) in one or several inner shells. The production of such multiple-electron holes is favoured by the considerable variety and multitude of atomic collision processes. The satellites of the $K_\alpha$ lines are of particular interest, which we describe as follows: We consider electron holes initially in the K- and L-shell for which the term symbols describing the holes are characterized by $-1$ for one-electron hole, $-2$ for two electron holes, $-3$ for three electron holes,.. ..., an initial distribution of holes may be transferred to a new distribution of holes, as indicated in this example.

$$K^{-1}L_I^{-1} \rightarrow L_{III}^{-1}L_I^{-1}, \alpha', 1S_{1/2}^{-1}2S_{1/2}^{-1}(1) \rightarrow 2P_{3/2}^{-1}2S_{1/2}^{-1}(2)$$

$$K^{-1}L_{II}^{-1} \rightarrow L_{III}^{-1}L_{II}^{-1}, \alpha''_0, 1S_{1/2}^{-1}2P_{1/2}^{-1}(1) \rightarrow 2P_{3/2}^{-1}2P_{1/2}^{-1}(2)$$

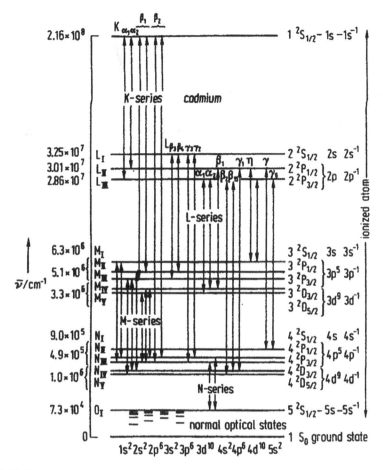

**Fig. 3.16** X-ray levels and X-ray spectral series of the cadmium atom. The normal optical levels, with the ground state as a zero point for the energies, are presented in the *lower part* of the figure. The exponent $-1$ in the configuration 1s, 2s, 2p indicates an electron hole in the sub-shells

$$\alpha_4 \left\{ \begin{array}{l} 1S_{1/2}{}^{-1}2P_{1/2}{}^{-1}(1) \rightarrow 2P_{3/2}{}^{-1}2P_{1/2}{}^{-1}(2) \\ 1S_{1/2}{}^{-1}2P_{1/2}{}^{-1}(0) \rightarrow 2P_{3/2}{}^{-1}2P_{1/2}{}^{-1}(2) \end{array} \right\}$$

$$K^{-1}L_{III}{}^{-1} \rightarrow L_{III}{}^{-1}L_{III}{}^{-1}, \left\{ \begin{array}{l} \alpha'_3, 1S_{1/2}{}^{-1}2P_{3/2}{}^{-1}(1) \rightarrow 2P_{3/2}{}^{-2}(2) \\ \alpha_3, 1S_{1/2}{}^{-1}2P_{3/2}{}^{-1}(2) \rightarrow 2P_{3/2}{}^{-2}(2) \\ \alpha''_3, 1S_{1/2}{}^{-1}2P_{3/2}{}^{-1}(1) \rightarrow 2P_{3/2}{}^{-2}(0) \end{array} \right\}.$$

**Fig. 3.17** Experimental data of the energy structure of K-absorption edges of free rare gas atoms (Ar and Kr) and their decomposition into components of various excite states (Breinig et al. 1980). The notations 4p, 5p, . . . cont. correspond to subsequent possible transitions $1s \rightarrow 4p$, $1s \rightarrow 5p$, . . ., $1s \rightarrow \infty$ of the K-absorption into optical states of Kr and Ar atoms. The decomposition into the 4p, 5p, . . .. states is based upon theoretical calculation of the above dipole transitions

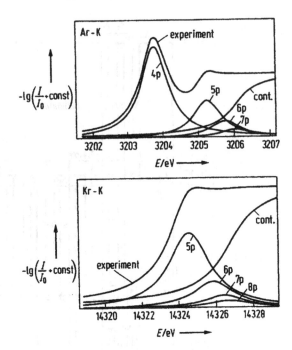

The figures in the round brackets represent the vector sum $j_1 + j_2$ of the joint electron holes. If double exchange of both electron holes is possible, we have additional transitions such as

$$K^{-1}L_I^{-1} \rightarrow L_{III}^{-1}L_{III}^{-1} \quad \text{or} \quad L_{II}^{-1}L_{III}^{-1},$$

$$\alpha'_0, 1S_{1/2}^{-1}2S_{1/2}^{-1}(0) \rightarrow 2P_{3/2}^{-1}2P_{1/2}^{-1}(1).$$

All the above satellite transitions are associated with the $K_{\alpha 1}$ line. A further number of transitions of double-electron holes may be expected with the $K_{\alpha 2}$ line ($K^{-1} \rightarrow L_{II}^{-1}$); however, such transitions have not yet been observed.

The satellite X-ray lines resulting from the redistribution of the electron holes are observed in the shorter wavelength part of the $K_{\alpha 1}$ line. This is due to the fact that the screening of the nuclear charge is reduced when the second electron is knocked out, since by then one of the electrons in the K-shell is already missing. Figure 3.18 shows the X-ray spectrum of the calcium $K_\alpha$ lines and their $K_{\alpha 1}$ satellite structure with the transition discussed above.

A large variety of satellite and hyper-satellite X-ray lines has been observed in high-energy collision between highly charged ions and atoms. A further interesting satellite effect is associated with two-electron-one-photon transitions. For example, if initially both electrons in the K-shell are missing, i.e. $1s^{-2}$, two transitions are possible, namely (1) $1s^{-1} \rightarrow 2p^{-2}$, which is an electric quadrupole transition E2

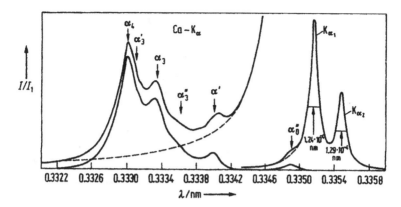

**Fig. 3.18** Calcium $K_\alpha$ X-ray lines including their satellite structure (*left-hand part*). The curves on the *right* represent the $K_{\alpha 1}$ and $K_{\alpha 2}$ lines. The intensity is represented in arbitrary units $I_1$, the wavelength is scaled in "X-ray units" XU, i.e. 1 kXU = 0.10024 nm (after Parrat 1936)

under the emission of one photon and (2) $1s^{-2} \to 2s^{-1} 2p^{-1}$, which is an electric dipole transition E1 with again an emission of one photon.

The energy of the photon associated with the transition of both electrons of the L-shell jumping into the K-shell is approximately twice as large, compared with "normal" $K_\alpha$ photons. It is usual to denote X-ray lines of simultaneous two-electron transitions by $K_{\alpha\alpha}$, $K_{\alpha\beta}$. .... lines depending on whether both electrons originate from the L-shell or one electron from the L-shell and the other from the M-shell in their transitions to the K-shell. Figure 3.19 shows the observation of $K_{\alpha\alpha}$ and $K_{\alpha\beta}$ lines produced by 3.5 MeV $Ar^+$ ions colliding with a target of calcium atoms. Based upon theoretical calculations and an analysis of the energy of the $K_{\alpha\alpha}$ lines as a function of the atomic number Z, it was concluded that the above E1 transition dominates between $Z = 12$ and $Z = 22$.

In line with our present concept of electron holes in inner shells, energy is released if an electron hole is refilled by an electron from a shell of smaller binding energy. In competition with the emission of a characteristic X-ray line in electron-hole transition, the following transition without radiation can take place. The energy released in a process of refilling an electron hole of a shell of higher binding energy may be "used" in knocking out an electron from a shell of lower binding energy. This process was discovered by the French physicist P. Auger in 1925 and is called the Auger effect (see Fig. 3.20). Let us assume that a projectile particle P (i.e. photons or atomic particles such as electrons, ions, ....) induces the ionization of an inner shell; the Auger effect may then be represented by the following reaction processes.

(1) $P + A \to P + A^+ + e(E_1) \to P + A^{++} + e(E_1) + e(E_{Auger})$
(2) $P + A \to P + A^{++} + e(E_1) + e(E_{Auger})$.

A represents the target atom, $e(E_1)$ is the knocked-out electron with the energy $E_1$ and $e(E_{Auger})$ is the Auger electron with sharp energy $E_{Auger}$. It has been found

**Fig. 3.19** X-ray spectrum produced in collisions between 3.5 MeV Ar$^+$ ions and calcium atoms. Various K lines of Ca and Ar are observed; $N$ is the number of the counted events (after Knudson et al. 1976)

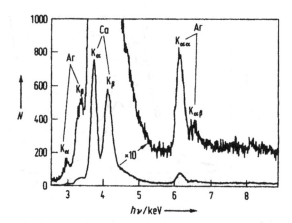

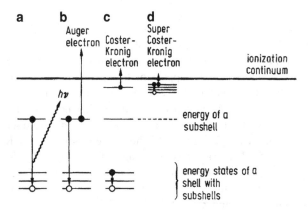

**Fig. 3.20** The various types of an electron hole (o) of inner shells of atoms (a) X-ray transition ($h\nu$), (b) Auger transition ($\uparrow$), (c) Coster-Kronig transition ($\uparrow$) and (d) super Coster-Kronig transition ($\uparrow$)

that the direct double-ionization process (2), which is a one-step process, has a much lower probability than the two-step process (1). It is common to characterize the Auger electrons (also called Auger transitions) in analogy to those of characteristic X-ray lines. The notations K-LL, or briefly KLL, means that initially, an electron is knocked out of the K-shell and subsequently two-electron holes in the L-shell are "produced" by the Auger effect. Correspondingly, KMM, LMM, …., Auger transitions are possible, whereby the sub-shell can be classified in addition by $KL_IL_{II}$, $KL_IL_{III}$ or by $KL_1L_2$, $KL_1L_3$. The double holes are also characterized by the coupling mechanisms such as $L, S, jj$ and intermediate couplings. However, if one of the second electron holes is in the same shell as the primary electron hole (but in another sub-shell), we talk about Coster-Kronig transitions as, for example, in $L_1L_2M$ or $M_1M_2N$ transitions which are special type of Auger processes (see

Fig. 3.20c). In general, the energies of Coster-Kronig electrons are smaller than those of Auger electrons. If both secondary electron holes are in the same shell as the secondary electron hole, we have a Super-Coster-Kronig transitions as, for example, the $L_1L_2L_3$ transitions (see Fig. 3.20d).

The Auger and Coster-Kronig electrons have sharp energies which result from the energy differences of the inner shells and sub-shells. Accordingly, the detection of Auger and Coster-Kronig transitions can be carried out by electron energy analyzers. A frequently applied energy analyzer is the cylindrical $127°$ analyzer as shown schematically in Fig. 3.21. Two cylindrically shaped metal plates are at positive and negative potentials, producing a radial electric field in the plane of the figure, the electric force $eE$ keeps the electron on circular trajectories if the centrifugal force is compensated, i.e. $E_e = mv^2/r$. It can be shown that electron passing through the entrance slit of analyzer under a small angle will be focused onto the exit slit. Varying the radial electric field strength changes the velocity $v$ or electron energy $E$ with which an electron can pass through the analyzer. This type of geometrical design can therefore be applied as a velocity or energy analyzer. Such $127°$ energy analyzer (or other types of analyzers for that matter) can be applied to detect Auger or Coster-Kronig electrons. Figures 3.22 and 3.23 show the typical examples of Auger and Costr-Kronig spectra.

We draw attention to an interesting effect in the Auger peaks which may occur by the ionization of inner shells through impact or photo-ionization, the effect is caused by a post-collision interaction (briefly called PCI effect) which takes place either by one outgoing electron (in photo-ionization) or by two outgoing electrons (in electron impact ionization) interacting with an Auger electron. This effect results in a small energy shift and broadening, and is strongest near the threshold energies since the outgoing electrons (or the two outgoing electrons) are still in the neighbourhood of the ion when the Auger electron is emitted.

According to the above, an electron hole can be refilled either by the process associated with the emission of a characteristic X-ray line or by Auger and Coster-Kronig transitions. These processes compete with each other; their relative contributions are called the fluorescence yield for the emission of characteristic X-radiation and Auger yield (including Coster-Kronig yields). Consider an initial situation in which $N_K$ atoms of a given kind have electron holes in their K-shell

**Fig. 3.21** Schematic arrangement of a spectrometer for the detection of Auger electron including a collision chamber and a $127°$ electron spectrometer. The Auger electrons are produced by electrons impinging on the atomic target in the centre of the collision chamber

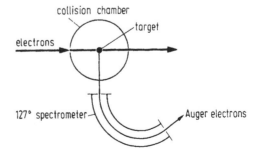

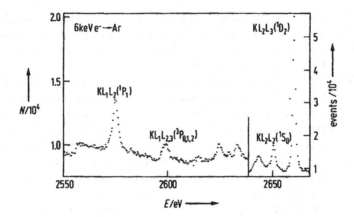

**Fig. 3.22** KLL Auger electron spectrum of argon atoms; the peaks not specifically assigned are Auger satellite lines, $N$ is the number of counted events (after Gräl and Fink 1985)

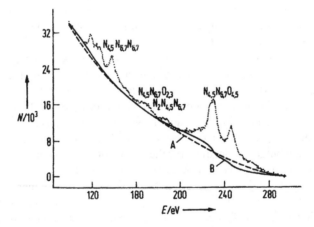

**Fig. 3.23** NNO Coster-Kronig and NNN super Coster-Kronig lines of mercury atoms produced by excitation with 3-keV electrons. The curves A and B describe theoretically the background signals of the electrons. $N$ is the number of events counted (after Aksela and Aksela 1983)

only; $N_{\gamma k}$ X-ray photons and $N_{AK}$ Auger electrons may then be emitted from these atoms. The relevant transition probabilities are denoted by $P_{\gamma K}$ and $P_{AK}$, respectively. We then define the fluorescence yield $\omega_k$ and the Auger yield $\alpha_K$ of the K-shell as follows:

$$\omega_K = \frac{N_{\gamma K}}{N_K} = \frac{P_{\gamma K}}{P_{\gamma K} + P_{AK}}, \quad \alpha_K = \frac{N_{AK}}{N_K} = \frac{P_{AK}}{P_{\gamma K} + P_{AK}} \tag{3.12}$$

with $\omega_K + \alpha_K = 1$.

For the L-shell, we have to take into account Coster-Kronig transitions in the decay of the electron holes; considering an initial population of $N_{L_i}$ ($i = 1, 2, 3$) for

L-electron holes, the Coster-Kronig yield can be defined by $f_{LL} = N_{LL}/N_L$ with $N_{LL}$ for the number of Coster-Kronig transitions $L_iL_KX$ ($X = M, N, O, \ldots$). In analogy to (3.12), we have the following branching for the $L_i$ sub-shells:

$$L_1-\text{shell:} \quad \omega_{L_1} + \alpha_{L_1} + f_{L_1L_1} + f_{L_1L_3} = 1,$$

$$L_2-\text{shell:} \quad \omega_{L_2} + \alpha_{L_2} + f_{L_2L_3} = 1,$$

$$L_3-\text{shell:} \quad \omega_{L_3} + \alpha_{L_3} = 1.$$

Fluorescence yields have been measured experimentally, particularly for K-, L- and M-shells; Fig. 3.24 shows some data (Bambaynek et al. 1972) where the following definitions are used for averaged fluorescence yields:

$$\bar{\omega}_{L_{2,3}} = \tfrac{1}{3}\omega_{L_2} + \tfrac{2}{3}\omega_{L_3} \quad \text{and} \quad \bar{\omega}_{M_{4,5}} = \tfrac{2}{5}\omega_{M_4} + \tfrac{3}{5}\omega_{M_5}.$$

The considerable increase of the fluorescence yield as a function of atomic number $Z$ is very striking and follows qualitatively from the consideration that the transition probabilities are proportional to $\gamma^3$ ($\gamma$ is the frequency of the X-ray line, which itself is proportional to $Z^2$). On the other hand, averaged radii in the squares of the dipole-matrix elements are proportional to $1/Z^2$, so that the fluorescence transition probabilities increase as $Z^4$. However, these estimates hold only for hydrogen-like systems to an approximation whose validity at high values of $Z$ is limited. Furthermore, it has been found that fluorescence yields also depend on the number of additional electron holes in higher shells as experimentally demonstrated particularly in ion-atom collisions.

## 3.1.11   Bremsstrahlung X-Rays

In addition to the characteristic X-ray line radiation, an additional continuous X-radiation is observed in the collision of charged particles with matter (Fig. 3.25). This type of X-radiation, which we have already noticed as a

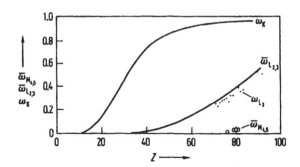

**Fig. 3.24** Experimental fluorescence yields as a function of the atomic number Z (after Bambaynek et al. 1972)

**Fig. 3.25** Continuous
Bremsstrahlung spectrum for
various energies of electrons
hitting tungsten as anode in an
X-ray tube. The intensity $I$ as
a function of the X-ray
wavelength $\lambda$ is given in
arbitrary units $I_1$ (after
Urey 1918)

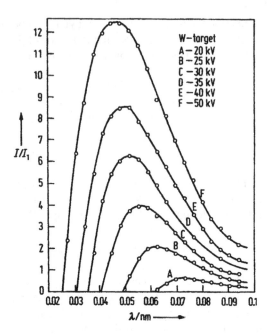

background radiation on which characteristic X-ray lines may appear (Fig. 3.26), is
due to the following physical process: electrons or other charged particles of energy
$E_o$ can lose kinetic energy in collisional interactions with atoms in a free or bound
state (i.e. in solids); a fraction of this energy loss can be transferred to electromagnetic radiation, which according to Sommerfeld is called Bremsstrahlung. The
collisional reaction for the production of Bremsstrahlung by electrons of energy
$E_o$ impinging on atoms A can be written as:

$$e(E_o) + A \rightarrow A + e(E_o - hv) + \gamma(hv).$$

The electrons lose energy, which is gained by the resulting X-ray photon $\gamma$ with
energy $hv$. The distribution of the energy $E_B = hv$ for Bremsstrahlung photons can
be calculated both classically and quantum mechanically. The main contribution in
the deceleration process of electrons impinging on atoms is due to interactions
with the orbital electrons. This causes excitation and ionization processes. If, however, the primarily incoming electrons completely or partially pass the electron
shells they can be deflected by the partially screened or pure Coulomb field of the
atomic nucleus. In 1923, the Dutch physicist Kramers assumed parabolic orbits for
the deflected electrons and calculated the energy and angular distribution of the
resultant continuous Bremsstrahlung, based upon the laws of classical electrody-
namics. The calculated angular distribution of the electromagnetic radiation corresponds to that of a classical oscillator vibrating in the direction of the incoming
electrons,

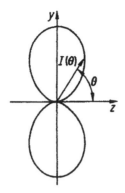

**Fig. 3.26** Angular distribution of the intensity $I(\Phi)$ of the Bremsstrahlung produced by electrons progressing parallel to the $z$-axis and striking an anode in the origin of the coordinate system (without relativistic correction). The radiation intensity is rotational symmetric around the $z$-axis and has its maximum in the equator plane ($9x - y$). It vanishes in the direction of deceleration ($z$-axis)

$$I(\theta) \sim \sin^2\theta; \tag{3.13}$$

where $\theta$ is the angle between the direction of the incoming electron and the direction of the observed Bremsstrahlung photons (Fig. 3.26). However, if the velocity $v$ of the incoming electron is comparable to the velocity of light $c$, a relativistic correction is required which modifies the angular distribution as follows (with $\beta = v/c$):

$$I(\theta) \sim \left[ \frac{1}{(1 - \beta\cos\theta)^4} - 1 \right] \frac{\sin^2\theta}{\cos\theta}. \tag{3.14}$$

Figure 3.27 shows how the dipole characteristic of (3.13) is modified by the relativistic correction included in (3.14). The emitted Bremsstrahlung radiation for high electron impact energies shows a preferred orientation around the forward direction with reference to the incoming electrons. Such relativistic effects have recently been studied in experiments with crossed electron and atomic beams. In these experiments with free atoms, the influence of solid bodies on the Bremsstrahlung is avoided. Figure 3.28 shows (see, e.g. Aydinol et al. 1980; Tseng et al. 1979) the relativistic asymmetry of the angular distribution of the Bremsstrahlung from free xenon atoms, even at a low electron energy ($E_e = 10$ keV).

The electrons which are decelerated and deflected in the Bremsstrahlung process transfer energy into electromagnetic radiation. According to Kramers, it can be predicted classically that the special distribution of Bremsstrahlung from free atoms or very thin targets of solids obeys the relation

$$I(Z, V, v)dv \sim \frac{Z^2}{eV} dv. \tag{3.15}$$

**Fig. 3.27** Relativistic effects in the angular distribution of the Bremsstrahlung intensity $I(\Phi)$ for various $\beta$ values ($\beta = v/c$) and kinetic energies $E_{kin}$ of the electrons hitting the target atoms (the z-axis is parallel to the direction of the incoming electrons)

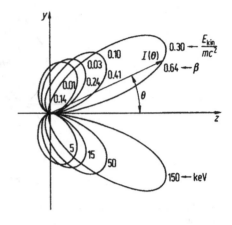

**Fig. 3.28** Relative angular distribution $I(\theta)/I(90°)$ of the Bremsstrahlung of free xenon atoms at an electron energy of $E_e = 10$ keV and an energy of Bremsstrahlung photons of $E_{ph} = 9.5$ keV. Experimental data (*closed circle*) after Aydinol et al. (1980); theoretical predictions (- - -) after Kuhlenkamp (1959) and (———) after Tseng et al. (1979)

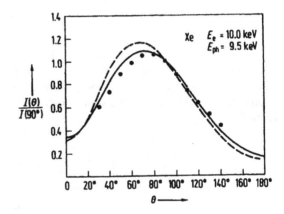

The spectral intensity of the Bremsstrahlung photons is proportional to the square of the atomic number Z, inversely proportional to the primary energy $eV$ of the incoming electrons but independent of the frequency or the energy of the Bremsstrahlung photons. This independence of the spectral or energy distribution of the Bremsstrahlung photons has indeed been verified for free atoms (Fig. 3.29) (Aydinol et al. 1980) and thin metal foils (Fig. 3.30). By considering thick solid targets as anodes, a superposition of many single-electron scattering processes take place from which the electron is more or less completely decelerated in the solid. Solid layers of equal thickness experience the same electron energy loss resulting in the same intensities of Bremsstrahlung per frequency interval. The Bremsstrahlung spectrum of solids can therefore be additively synthesized from the Bremsstrahlung spectra of atoms (Fig. 3.31). This is in agreement with the experimentally observed Bremsstrahlung spectra of solids for which the intensity decreases linearly with the frequency or energy of the photons (Fig. 3.32) and which is drastically different from the Bremsstrahlung spectrum of free atoms (Fig. 3.29; Kuhlenkempf and Schmidt 1943).

In this connection, we would like to draw attention to the limited validity of the classical theory of Bremsstrahlung. More complicated quantum-mechanical calculations are necessary to take into account correctly the deviations from the classical $Z^2$ dependence of Bremsstrahlung as shown in Fig. 3.33 (Hippler et al. 1981; Pratt et al. 1977).

Finally, we refer to recent research on Bremsstrahlung in which electron-spin methods and coincidence detection for the inelastically scattered electrons and the Bremsstrahlung photons are being applied (Nakel 2006).

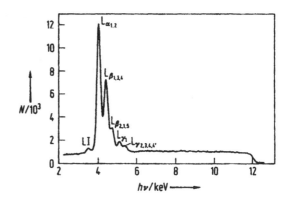

**Fig. 3.29** Experimental data of the X-ray spectrum of free xenon atoms for excitation by 12 keV electrons. The contribution of the Bremsstrahlung is represented by horizontal recorder curve while the peaks are characteristic X-ray lines. $N$ is the number of the counted events (after Aydinol et al. 1980)

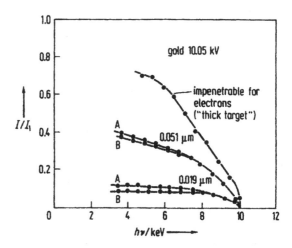

**Fig. 3.30** Intensities (in relative units of $I_1$) of the Bremsstrahlung spectra of gold for excitation by 10.05 keV electrons at various thicknesses of foils and thick target; curves B are corrected for a contribution from the Bremsstrahlung produced in the carrier foil while curves A are uncorrected

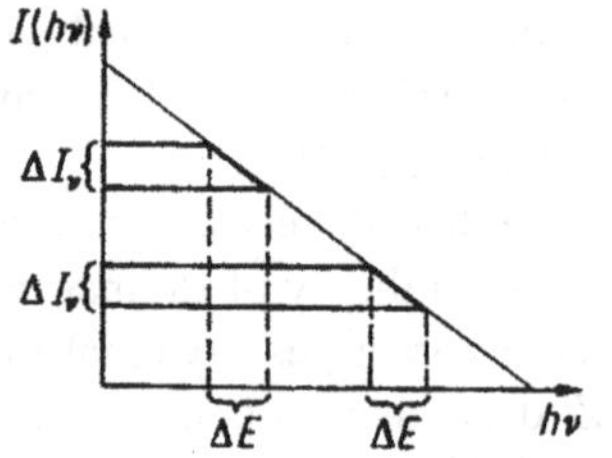

**Fig. 3.31** Build-up of a Bremsstrahlung spectrum $I$ ($h\nu$) for which layers of equal energy loss $\Delta E$ of the electrons result in equal intensity contributions $\Delta I$ per frequency interval

**Fig. 3.32** Intensities $I$ (in relative units of $I_1$) of Bremsstrahlung spectra of a thick platinum target for primary electron energies from 49 to 19.8 keV (after Kuhlenkempf and Schmidt 1943)

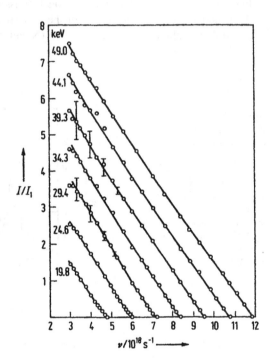

## 3.1.12   Crystal X-Ray Spectrometer

### 3.1.12.1   The Principle of Crystal X-Ray Spectrometer

Bragg's law $n\lambda = 2d \sin \theta$ relates the wavelength $\lambda$ of the incident X-ray to the angle $\theta$ at which a coherent diffraction pattern is formed. Here, $d$ is the crystal lattice spacing and $n$ is an integer which gives the order of reflection. A simple spectrometer uses a planar crystal, Soller slits (see Sect. 3.1.12.3) for X-ray collimation and a proportional counter (see Sect. 3.1.12.5) for X-ray detection.

The angular dispersion $D_\theta$ for a Bragg's spectrometer is given by

**Fig. 3.33** Bremsstrahlung intensity as a function of the atomic number Z; the primary energy of the electrons colliding with the atoms is $E_0 = 2.5$ keV; the energy of the Bremsstrahlung photons is $E_{ph} = 2.0$ keV. Experimental data points after Hippler et al. (1981)

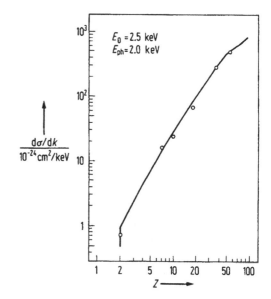

$$D_\theta = \frac{d\theta}{d\lambda} = \frac{\tan\theta}{\lambda} = \frac{n}{2d}\sec\theta. \tag{3.16}$$

From (3.16) it can be seen that high dispersion is obtained for larger $\theta$, or at backward angles. Also for a given crystal, better resolution can be obtained, in general, by measuring the X-rays in a higher order.

The energy resolution of a crystal spectrometer is obtained from the derivation for $d\theta/d\lambda$ and can be expressed as

$$\frac{\Delta E}{E} = \cot\theta\Delta\theta. \tag{3.17}$$

The angular resolution $\Delta\theta$ is determined by the instrumental resolution and by the angular spread (usually less than $0.1°$) corresponding to the rocking curve (Werner 1983) of the crystal. The rocking curve is a measure of the angular range over which diffracted monochromatic X-rays are spread due to crystal imperfections. It follows from (3.17) that the energy resolution worsens with increasing X-ray energy. Thus, while crystal spectrometers provide an excellent resolving power ($\sim 10^3$) at low photon energy, they become inferior to energy dispersive solid-state devises (Wille and Hippler 1986) for energies above 20 keV.

## 3.1.13   The Crystal X-Ray Spectrometer

Figure 3.34 shows schematic diagram of the spectrometer. M is a three phase stepping motor (Sigma Instruments) which rotates a shaft S by the help of a flexible

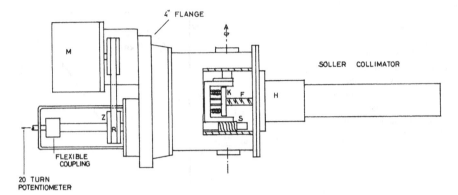

**Fig. 3.34** The crystal X-ray spectrometer. K is the crystal; F is the proportional counter; S is the shaft responsible for rotating the crystal; H is the holder for the Soller collimator; M is the stepping motor and R is the flexible belt

belt R. The shaft S rotates the crystal K, mounted on a plastic holder, around the axis of rotation A, in steps of $\sim 0.01°$. A Soller collimator, held in place by a holder H, collimates the beam of X-rays to an accuracy of $0.075°$. The reflected X-rays enter the sensitive volume of a proportional counter through its thin window F. A 20-turn potentiometer is fixed to the shaft S by a flexible coupling for locating the position of the crystal.

### 3.1.14   The Collimation of X-Rays

X-rays are collimated by passing through a Soller-slit-system (Siemens AG). This is an arrangement of plane parallel metal plates uniformly separated to allow a well-defined parallel beam of X-rays. The divergence of the beam of X-rays is then given by

$$\tan \varepsilon = \frac{S}{\ell},$$

where $d$ is the thickness of the Soller plates, $S$ the distance between them and $\ell$ is their length as shown in Fig. 3.35.

A good collimation ($2\varepsilon = 0.075°$) has been attained with a Soller collimator (Siemens AG) using plates having $\ell = 150$ mm, $S = 0.10$ mm and $d = 0.10$ mm. An analysis of X-ray transmission by various Soller collimators has been carried out by Harbach (1980).

### 3.1.15   The Crystal

Incident X-rays are reflected specularly by a crystal as shown in Fig. 3.36. The reflectance for a ray incident at a glancing angle $\theta$ is denoted by $CW[\theta - \theta_{\rm B}(\lambda)]$

**Fig. 3.35** The Soller collimator. $\ell$ is the length of the Soller plates; $d$ is the thickness of the plates and $S$ is the separation between the successive plates

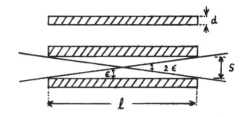

**Fig. 3.36** Reflection of X-rays by a plane crystal

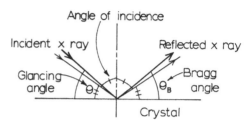

(Morita 1983) where $C$ is a constant, $W[\theta - \theta_B(\lambda)]$ is a function whose graph is called the rocking curve and $\theta_B(\lambda)$ is the Bragg angle, which satisfies the Bragg's law

$$\{2d \sin \theta_B(\lambda) = n\lambda\}.$$

Here, $\lambda$ is the wavelength of the incident ray, $d$ is the spacing of the crystal lattice and $n$ is the order of the diffraction.

If a collimated beam of X-rays is incident on a crystal then the intensity distribution of an X-ray spectrum can be measured by a detector placed at the corresponding angle on the opposite side of the normal. The system effectively selects all rays which are incident at the appropriate angle of Bragg reflection. If the angle of incidence is then varied (by rotating the crystal through a small angle), the detector will then receive the radiation of a different wavelength and thus a spectrum will be obtained.

The resolution of such a spectrometer depends, in part, on the geometrical definition provided by the slit. A further limit is imposed by the intrinsic resolution of the crystal. If the incident radiation is truly monoenergetic, the peak obtained in the spectrum should be that for the crystal alone. In practice, however, the characteristic X-ray lines are found to have a finite width, which is related to the lifetime of the state by the uncertainty principle. The natural widths of the X-ray lines can be studied by using spectrometers having high resolution.

The natural resolution of crystals depends upon the degree of alignment of the surface atomic planes. For mosaic crystals to be suitable for spectroscopy, they

**Table 3.3** Analyzing crystals (Jenkins and de Vries 1970)

| Crystal | Reflection plane | 2d Spacing (Å) | Lowest atomic number detectable | | Reflection efficiency |
|---|---|---|---|---|---|
| | | | K-series | L-series | |
| Opaz | (303) | 2.712 | V (23) | Ce (58) | Average |
| Lithium fluoride | (220) | 2.848 | V (23) | Ce (58) | High |
| Lithium fluoride | (200) | 4.028 | K (19) | In (49) | Intense |
| Sodium chloride | (200) | 5.639 | S (16) | Ru (44) | High |
| Quartz | (10$\bar{1}$1) | 6.686 | P (15) | Zr (40) | High |
| Quartz | (10$\bar{1}$0) | 8.50 | Si (14) | Rb (37) | Average |
| Penta erythritol | (002) | 8.742 | Al (13) | Rb (37) | High |
| Ethylenediamine tartrate | (020) | 8.808 | Al (13) | Br (35) | Average |
| Ammonium dihydrogen phosphate | (110) | 10.65 | Mg (12) | As (23) | Low |
| Gypsum | (020) | 15.19 | Na (11) | Cu (29) | Average |
| Mica | (002) | 19.8 | F (9) | Fe (26) | Low |
| Potassium hydrogen phthalate | (10$\bar{1}$1) | 26.4 | O (8) | V (23) | Average |
| Lead stearate | | 100 | B (5) | Ca (20) | Average |

should have a narrow rocking curve combined with a high integrated reflection coefficient [the reflection coefficient integrated over the angular width comprising the rocking curve (Werner 1983; Birks 1978)].

A list of analyzing crystals is given in Table 3.3, showing their main characteristics together with the lowest atomic number detectable in each case.

The double spectrometer is widely used for the examination of the line profiles at ordinary wavelengths because the dispersion available is twice that of a single crystal spectrometer and it also has a much better resolution.

## 3.1.16  Proportional Counter

The proportional counter is a special form of ionization chamber in which a single ionization event produces an output pulse which has been amplified $10^4$–$10^7$ times (Fink 1975), due to electron multiplication in the high-electric-field region surrounding the central anode wire. If there is a strong electric field to give the free electrons enough energy, between collisions with the gas molecules, to cause additional ionization then a multiplication of the free electrons can occur.

In a coaxial cylindrical counter having an anode wire of radius "$a$" and a cylindrical cathode of radius "$b$," the electric field $E$ at a distance "$r$" from the centre is given by

$$E = \frac{V}{[r \log e \ (b/a)]},$$ (3.18)

where $V$ is the total voltage applied to the counter. This shows that the region where an electron gains sufficient energy between collisions to cause further ionization lies within a region of a few diameters of the anode wire, so that the output pulse height is essentially independent of the location of the initial ionizing event in the counter volume.

The gas amplification $M$ can be defined as the number of electrons collected at the central wire per primary electron released in the original ionizing event. The specific ionization $\alpha(r)$ is the mean number of secondary electrons produced by an electron per unit centimetre of its path. The specific ionization is taken as the reciprocal of the mean free path between ionizing collisions which is a function of the radial distance $r$ from the centre. Gas amplification and specific ionization are related by

$$\text{Log } M = \int_a^{r_C} \alpha(r) \, dr, \tag{3.19}$$

where $r_C$ is the critical radius where the multiplication of the free electrons first begins.

It can be shown (Gold and Bennet 1966) that the energy resolution of a proportional counter is determined largely by the gas amplification, which in turn depends critically on the counter wire radius. Also, the quality of the central wire, i.e. its smoothness and uniformity, is important for good resolution.

The energy resolution $\Delta E$ of a proportional counter is to a large extent determined by the statistics of the ionization processes taking place in the counter gas. For the X-ray peak with energy $E$, the energy resolution defined as full width at half maximum (FWHM) due to collection statistics is given by Mokler and Folkmann (1978). A typical value of the energy resolution of a proportional counter is about 800 at an X-ray energy of 6 keV (Mokler and Folkmann 1978).

It has been observed (Culhane et al. 1966) that the rise-time of the pulses originating from X-rays is rather faster than due to high-energy electrons and mesons. This appears to be due to the short range of photo-electrons with the consequence that all the ion-pairs produced by the absorption of an X-ray photon originate in a closely defined region of the counter, causing the electrons to have transit times which are closely similar. The differences in rise-time have been used (Campbell 1968; Lewyn 1970; Isozumi and Isozumi 1971) in the X-ray region of 6–15 keV energy, for example, to reject more than 90% of the background while more than 95% of the X-ray signals are retained by the system. Rise-time discrimination of X-rays from particle events is discussed in an article by Culhane and Fabian (1972).

The proportional counter used has an active length of $\sim 7$ cm, cathode diameter 2.1 cm and tungsten wire of diameter 0.1 mm. The operating potential is 1,500 V.

The experimental techniques are given in Sects. 4.1–4.3, the results and discussion are described in Sects. 5.1–5.1.6 and the concluding remarks are given in Chap. 6.

## 3.2    The Apparatus for the Electron-Molecule Collision Process

### 3.2.1    The Apparatus and the Electronic Equipment Used in the Experiment

Figure 3.2 shows the apparatus inside the vacuum chamber. The electron gun (right), the Faraday cup (left), the ion analyzer (top) and the electron analyzer (at the back) are all supported on turntables to facilitate their independent motion. The electron gun is described in Sect. 3.1.3 and is shown schematically in Fig. 3.3. The molecular beam source is described in Sect. 3.1.4 and the 30° parallel plate electron analyzer, shown in Fig. 3.5, is described in Sect. 3.1.5. A TOF type ion analyzer, used in this experiment to analyze the produced ions, is described in Sect. 3.1.6 and is shown in Fig. 3.7.

The CAMAC system, used in this experiment for data acquisition, is a programmable data handling system controlled by a micro-computer and it consists of an instrument mainframe (crate), CAMAC to GPIB interface and a Time-to-Digital Converter (TDC). LeCroy (model 8013A) crate is providing the housing and power for the TDC and GPIB interface and accepts instruments of up to 13 single width plug-ins. The crate contains all the data-way lines between the TDC and the computer via CAMAC to GPIB interface which control the CAMAC system and operates as a "talker-listener." All communication takes place through the data-way lines and this includes the digital data transfers, the strobe signals, the addressing signals and the control signals.

The LeCroy model 4208 multi-hit TDC with a resolution of 1 ns is the heart of the CAMAC system and records the arrival times of the produced ions with reference to the detected electron. The CAMAC module (LeCroy model 8901A) providing access between the CAMAC mainframe and the computer, which uses GPIB (IEEE 488) interface, makes it possible to interface and control modular instruments in the crate. There are also other research grade electronic instruments like signal amplifiers and discriminators which are being used in the experiment.

The experimental technique for this experiment is given in Sect. 4.4, the results and discussion are described in Sect. 5.2 and the conclusion is given in Chap. 6.

## 3.3    Apparatus for the Study of the Excitation of Spin-Polarized Atoms of Sodium and Potassium

### 3.3.1    Vacuum Chambers for Housing the Apparatus

Figure 3.37 shows a photograph of the apparatus. It consists of two large and one small vacuum chamber. One of the large vacuum chambers accommodates the oven where alkali metals are heated to produce the atomic beam and the other large

**Fig. 3.37** A photograph of the apparatus

chamber is used to house the interaction region. The small chamber, which is placed in between the two large chambers, houses the hexapole magnet for polarizing the atomic beam. The collision chamber has one whole side formed by a big flange on which a plate is mounted which supports a turntable 670 mm in diameter. The big flange is bolted to a trolley and can be put away to allow easy access to all the components mounted on the turntable. A long vacuum tube is connected to the centre of the big flange and near the end of this tube is mounted a Rabi magnet followed by a Langmuir-Taylor detector. The experiment is usually carried out at $\sim 6 \times 10^{-7}$ mbar.

## 3.3.2 Cancellation of the Earth's Magnetic Field

After cancellation of the magnetic field in the collision region, the measured magnetic field is $\sim 10$ mG there and $\sim 35$ mG near the electron gun.

## 3.3.3 The Components of the Atomic Beam Apparatus

### 3.3.3.1 The Oven for the Vapourization of Alkali Metals

The oven (shown in Figs. 3.38 and 3.39) is designed to vapourize alkali metals. The oven is made from molybdenum with a nozzle aperture of 1.5 mm diameter. The main body of the oven can hold 50–60 g of the alkali metal. Three screws at

**Fig. 3.38** The oven assembly inside the vacuum chamber

the bottom of the oven allow the adjustment of the oven in any position and help in preventing the heating of the oven-holding-plate at the bottom. The oven as a whole can be moved forward or backward to enable its placement at a desired distance from the hexapole magnet, in front. The main body of the oven is shielded with a stainless steel foil wrapped around it. The temperatures of the main body and the nozzle are measured with two thermometers placed in contact with them. The main body of the oven is heated with six 5 W, resistors of 50 $\Omega$ resistance inserted into holes drilled from bottom into the wall of the main body, while the interconnecting tube and the nozzle are heated with thermo-coaxial wires.

The heating of the oven is started by first heating the interconnecting tube and the nozzle to clear any residual metal and then the main body is heated and a temperature difference of 30–50°C is always maintained between the nozzle and the main body. For sodium, the operating temperature is 350°C for the main body and 390°C for the nozzle while for potassium the main body and the nozzle are kept at 270 and 300°C, respectively. This temperature difference between the main body of the oven and the nozzle prevents any clogging of the nozzle. A water-cooled condenser is placed in front of the nozzle of the oven to collect the metal vapours going in other directions than the beam.

### 3.3.3.2 Dimer Reduction and Atomic Beam Intensity

The oven has two stages, the main body and the nozzle, which are connected via a tube called the interconnecting tube. The nozzle consists of an aperture in the thin

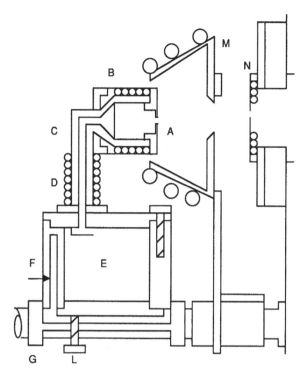

**Fig. 3.39** Schematic diagram of the oven. B is the oven nozzle with aperture A; C is the interconnecting tube; D is a heated coaxial tube; E is the main body of the oven; F is the heater element for the oven; G is the supporting base for the oven; L is a screw for adjusting the position of the oven; M is a water-cooled condenser for condensing the vapours which are not in the beam and N is a heated aperture

wall of the oven through which the alkali metal vapours effuse to the surroundings. Assuming that the laws of simple effusion from an aperture in a thin wall and from a long channel hold then, following Ramsey (1969) and Melisse and Moody (1977), if we let the condition of the main body be denoted by subscript 2 and the nozzle by subscript 1; indicate the values of density and temperature in such regions by the symbols $n$ and $T$, respectively; denote by $A_{can}$ the cross-sectional area of the canal of the interconnecting tube and by $A_{nzz}$ the area of cross-section of the aperture of the nozzle; $I_{\uparrow}$ and $I_{\downarrow}$ are the flow up and down the interconnecting tube and $I_0$ is the rate of effusion from the output aperture of the nozzle, then

$$I_0 = \left(\frac{1}{4}\right) n_1 \underline{v_1} A_{nzz}$$

$$I_{\downarrow} = \left(\frac{1}{p}\right)\left(\frac{1}{4}\right) n_1 \underline{v_1} A_{can} \,.$$  (3.20)

$$I_{\uparrow} = \left(\frac{1}{p}\right)\left(\frac{1}{4}\right) n_2 \underline{v_2} A_{can}$$

Factor $1/p$ in the above equations corrects the fact that the connecting tube is not a thin aperture and its value is given by

$$\frac{1}{p} = \left(\frac{8}{3}\right)\left(\frac{r}{L}\right) = \frac{1}{8},\tag{3.21}$$

where $r$ is the radius of the canal of the interconnecting tube (0.25 cm) and $L$ is the length of the interconnecting tube (5.5 cm).

In equilibrium, the condition that must be met is

$$I_0 = I_\uparrow - I_\downarrow \tag{3.22}$$

From (3.20) and (3.22), we get

$$\frac{n_1}{n_2} = \frac{v_2}{v_1} \frac{1}{[P(A_{nzz}/A_{can}) + 1]}\tag{3.23}$$

since $v = (8kT/\pi M)^{1/2}$, therefore,

$$\frac{v_2}{v_1} = \left(\frac{T_2}{T_1}\right)^{1/2},\tag{3.24}$$

where $T_2$ is the temperature of the main body and $T_1$ is the temperature of the nozzle.

For this oven $A_{nzz}/A_{can} = (5/3)^2$, therefore, (3.23) becomes

$$\frac{n_1}{n_2} = 0.446\left(\frac{T_2}{T_1}\right)^{1/2}.\tag{3.25}$$

For sodium atoms, $T_2 = 623$ K and $T_1 = 663$ K, therefore,

$$\frac{n_1}{n_2} = 0.419.\tag{3.26}$$

For potassium atoms, $T_2 = 538$ K and $T_1 = 573$ K, thus

$$\frac{n_1}{n_2} = 0.419.\tag{3.27}$$

We find that, by keeping the nozzle at higher temperature than the main body of the oven, not only the density of the alkali metal in the nozzle region is half of that in the main body, but also there is a reduction in the dimer density (Melisse and Moody 1977).

By using the estimation method of Ramsey (1969), we find the intensity of the atomic beam as follows:

If $N(\theta)$ $d\Omega$ is the number of particles effusing per second at an angle $\theta$ with respect to the normal of the area $A_{nzz}$ of the source then

$$N(\theta) \; d\Omega = \frac{n\underline{v}A_{nzz}}{4\pi} \cos\theta d\Omega, \tag{3.28}$$

where $n$ is the number of particles per unit volume and the average velocity of the atom is given by

$$\underline{v} = \left\{\frac{8kT}{\pi m}\right\}^{1/2},$$

where $T$ denotes the temperature, $m$ is the mass of the particle and $k$ is the Boltzman's constant.

For sodium $\underline{v} = 7.812 \times 10^4$ cm/s and for potassium $\underline{v} = 5.572 \times 10^4$ cm/s. $A_{nzz}$, the area of cross-section of the nozzle, $= 0.018$ cm$^2$. $n$, the atomic density is given by $n = p/kT$, where $T$ is the temperature, $p$ is the pressure and $k$ is the Boltzman's constant.

From Fig. 3.40, showing vapour pressure/temperature relation for sodium and potassium, we get the pressure for sodium, $P_{Na} = 1 \times 10^{-1}$ mmHg at 623 K (main body) and for potassium, $P_K = 9 \times 10^{-2}$ mmHg at 538 K (main body).

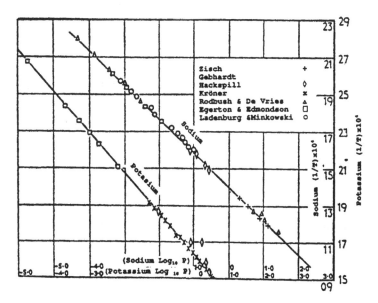

**Fig. 3.40** Vapour pressure versus temperature graphs for sodium and potassium. $T$ is the temperature in K and $P$ is the vapour pressure in mm of Hg

Substituting these values in the equation $n = p/kT$ we get

for sodium, $n_{Na} = 1.55 \times 10^{15}$ atoms/cm$^3$ (main body) and

for potassium, $n_K = 1.63 \times 10^{15}$ atoms/cm$^3$ (main body).

Equation (3.25), which relates the density $n_1$ in the nozzle and the density $n_2$ in the main body, gives

for sodium, $n_{Na}$ (at nozzle) $= 0.67 \times 10^{15}$ atoms/cm$^3$ and

for potassium, $n_K = 0.7 \times 10^{15}$ atoms/cm$^3$.

Integrating (3.28), we get

$$N = \frac{n v A_{nzz}}{4} \tag{3.29}$$

and substituting the values, we get

for sodium, $N = 2.7 \times 10^{17}$ atoms/s and $\tag{3.30}$

for potassium, $N = 1.8 \times 10^{17}$ atoms/s. $\tag{3.31}$

The maximum intensity of the beam in the forward direction which can be transmitted through the hexapole magnet is (Hils 1971)

$$I_{\pm} = \left(\frac{N}{\pi}\right)\Omega_{\pm}, \tag{3.32}$$

where ($\pm$) refers to the spin up and down components of the beam and ($\Omega_{\pm}$) is the solid angle of acceptance of the hexapole magnet for the two spin states (the value of which is given in the next section).

### 3.3.3.3  The Hexapole Magnet

Hexapole (six-pole) magnet, first introduced by Friedburg (1951), can be used to separate particles having different spin states (Hughes et al. 1972) and to focus the particles of one spin state and defocus the particles of opposite spin state. The hexapole magnet has a cylindrical shape and the magnetic potential at the wall of the magnet varies as $\sin(n\theta)$, where $\theta$ is the azimuthal angle, and $n$ is a positive integer.

For a magnet, having $N$ pairs of pole tips with each tip occupying an angle $\alpha$ on the cylinder of radius $r_0$, if the magnetic potential varies from one tip of the magnet to the other tip as $\pm V$ and also if the potential is zero at the centre of the magnet then the magnetic potential $\Phi(r, \theta)$ inside the magnet is given by

$$\Phi(r,\theta) = \frac{4V}{\pi} \sum_{n=1}^{\infty} \left(\frac{r}{r_0}\right)^n \frac{\sin n\theta}{n} \sin \frac{n\alpha}{2} \sum_{L=1}^{N} (-1)^{L-1} \times \sin\left[\frac{n}{N} \frac{\pi}{2} (2l - 1)\right]. \quad (3.33)$$

Since the magnetic field $H = -\nabla\Phi$, the (3.33) leads to

$$H = H_0 \left(\frac{r}{r_0}\right)^2 \left[1 - 2\left(\frac{r}{r_0}\right)^6 \cos 6\theta + \left(1 - 2\cos 12\theta \left(\frac{r}{r_0}\right)^{12} + \ldots\right)\right]^{1/2}, \quad (3.34)$$

where $r$ is the radial distance from the hexapole axis, $r_0$ is the radius of the hexapole magnet and $H_0$ is the magnetic field at the tip of the magnetic poles and is measured by a Tesla meter as 8,600 G.

The hexapole magnet being used in this experiment (Fig. 3.41) has $N = 3$ pairs, a length of 7.6 cm, radius $r_0 = 0.159$ cm and each pole-tip occupies an angle of 30°. The magnetic field $H$ is given by

$$H = H_0 \left(\frac{r}{r_0}\right)^2. \quad (3.35)$$

For an atom, having a magnetic moment $\mu$, the energy $E$ in a magnetic field $H$ is given by

$$E = -\mu . H \quad (3.36)$$

and this equation holds when $\mu$ is constant and parallel to an associated angular momentum and any change in $H$ is applied slowly (in many cycles of Larmor precession of the moment in the field), then the projection of $\mu$ on $H$ remains constant (i.e. there is no change in quantum state). Further, that the direction of the magnetic field $H$ has no effect on the energy $E$ of the system. The magnetic field inside the hexapole magnet varies from a weak field due to the penetration of the earth's magnetic field at the axis of the magnet to a strong magnetic field at the tips of the hexapole. The force acting on the atom is given by

$$F = -\nabla_r E = -\left(\frac{\delta E}{\delta H}\right) \nabla |H| \quad (3.37)$$

and putting $(\delta E / \delta H) = -\mu_{\text{eff}}$ we get

$$F = 2\mu_{\text{eff}} H_0 \left(\frac{r}{r_0^2}\right). \quad (3.38)$$

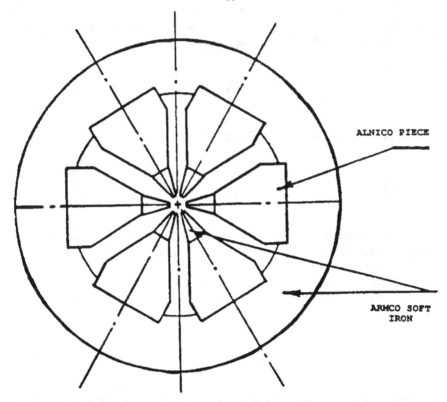

ALNICO PIECE

ARMCO SOFT
IRON

**Fig. 3.41** Schematic diagram of a hexapole permanent magnet. The cylindrical yoke and the pole tips are made from Armco soft iron; pyramid-shaped poles are made from Alnico V

Figure 3.42 shows (Ramsey 1969) the variation of the effective magnetic moment of the atom, having nuclear spin $I = 3/2$, with the magnetic field. For higher magnetic field

$$\mu_{\text{eff}} = -m_J g_J \mu_0 \, (\text{Christensen} 1959) \tag{3.39}$$

$$\approx \pm \mu_0$$

where $\mu_0$ = Bohr magnetron.

Substituting these values in (3.38), we get, for atoms with $\delta E/\delta H > 0$ and, therefore, $m_J > 0$, a force directed towards the axis of the hexapole magnet

$$F_+ = -\frac{2\mu_0 H_0}{r_0^2} r \tag{3.40}$$

**Fig. 3.42** Variation of the effective magnetic moment with the magnetic field. The *dotted lines* indicate the moments of the magnetic levels arising from the $F = I - 1/2$ state. The nuclear moment is assumed to be positive $X = ((\mu_J/J)H)/\Delta E$

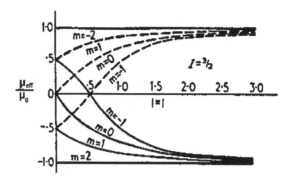

and for atoms with $\delta E/\delta H < 0$ and, therefore, $m_J < 0$, a force directed away from the axis of the hexapole magnet

$$F_- = +\frac{2\mu_0 H_0}{r_0^2}r. \tag{3.41}$$

The atoms with the force $F_+$, directed towards the axis of the hexapole, are those which get focused and are the ones used for the experiment while those atoms which are acted upon by a force $F_-$ move away from the axis and are defocused and get out of the atomic beam. This effect is more evident from the Breit-Rabi diagram for a spin 1/2 particle with a nuclear spin 3/2 shown in Fig. 3.43 (Ramsey 1969). Figure 3.44 shows the focused atomic beam as it exits the hexapole magnet.

To find the intensity of the focused atomic beam as it leaves the hexapole magnet, let us consider that the hexapole entrance subtends a solid angle $\Omega$ at the nozzle aperture which is located at a distance $L_1$ away from the entrance of the hexapole and if the initial velocity of the atoms making a small angle $\phi_0$ with the hexapole axis is $v$ then for a distance $x$ measured within the magnet the path of the trajectory, for atoms having quantum number $m_J > 0$, is given by (Christensen 1959)

$$r(x) = \varphi_0\left[L_1 \cos\left(\frac{\omega x}{v}\right) + \left(\frac{v}{\omega}\right)\sin\left(\frac{\omega x}{v}\right)\right]. \tag{3.42}$$

and for atoms, with quantum number $m_J < 0$, the equation of the trajectory is

$$r(x) = \varphi_0\left[L_1 \cosh\left(\frac{\omega x}{v}\right) + \left(\frac{v}{\omega}\right)\sinh\left(\frac{\omega x}{v}\right)\right], \tag{3.43}$$

where $L_1 = 11$ cm and $\omega = (2\mu_{\text{eff}}H_0)^2/(mr_0^2)$, but for a $^2S_{1/2}$ state and a large value of $H_0$

$$\omega = \frac{(2\mu_0 H_0)^2}{mr_0^2} \tag{3.44}$$

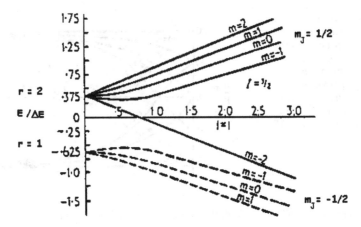

**Fig. 3.43** Variation of the energy with the magnetic field. Nuclear moment is assumed to be positive, $J = 1/2$. The *dotted lines* show the magnetic levels arising from the $F = I - 1/2$ state. The level structure corresponds to the ground state of sodium and potassium. $\Delta E$ is the energy difference between $F = 2$ and $F = 1$ at zero magnetic field

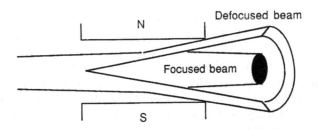

**Fig. 3.44** The focused and the defocused components of the atomic beam after passing through a hexapole magnet

and $\mu_o$ is equal to the Bohr magnetron.

For hexapole magnet, the solid angle of acceptance for the spin-up atoms is given by

$$\frac{\Omega_+}{4\pi} = \frac{\varphi_o^2}{4} = \frac{1}{4} \cdot \frac{r_o^2}{L_1^2 + (v/\omega)^2} \tag{3.45}$$

and that for spin down atoms by

$$\frac{\Omega_-}{4\pi} = \frac{1}{4} \cdot \frac{r_o^2}{[L_1 \cosh(\omega L_m/v) + (v/\omega)\ \sinh(\omega L_m/v)]^2}, \tag{3.46}$$

where $L_m = 7.6$ cm is the length of the hexapole magnet.

To find the intensity of the spin-up atoms, we need to calculate $\Omega_+$ using (3.45)

$$\Omega_+ = \cdot \frac{\pi r_0^2}{L_1^2 + (v/\omega)^2} \cdot \tag{3.47}$$

Since $r(x)$ and $\Omega$ depend on the velocity of the atoms, the value of $\Omega_+$ should be averaged over the velocity distribution, i.e.

$$<\Omega_+>_v = \frac{\int_0^\infty f(v)\Omega_+ dv}{\int_0^\infty f(v)dv}, \tag{3.48}$$

where $f(v) = (2I_0/\alpha^4)v^3 \exp(-v^2/\alpha^2)$ (Ramsey 1969), $\alpha = (2kt/m)^{1/2}$ is the most probable velocity and $I_0 = \int_0^\infty f(v)dv$ is the total intensity.

For sodium, at 663 K oven temperature, calculations show that

$$<\Omega_+>_v = 4.234 \times 10^{-4} st. \tag{3.49}$$

and for potassium, at 573 K oven temperature

$$<\Omega_+>_v = 4.9 \times 10^{-4} st. \tag{3.50}$$

Equation (3.32) gives the intensity of the focused beam as

$$I_+ = \left(\frac{N}{\pi}\right)<\Omega_+>_v.$$

Therefore, the intensity of the atomic beam for sodium is

$$I_{+(Na)} = 3.6 \times 10^{13} \text{ atoms/s}$$

and for potassium

$$I_{+(K)} = 2.8 \times 10^{13} \text{ atoms/s}.$$

### 3.3.4 Guiding Fields and the Low Field Polarization of Atoms

In the previous section, it was mentioned that for an atom having magnetic moment $\mu$, fixed in magnitude and linked to an associated angular momentum, there would be no change of state of the atom if there is a slow variation (i.e. in many cycles of Larmor precession of the magnetic moment in the field) in the magnetic field $H$.

This condition is met inside the hexapole magnet but problems can arise when the atoms leave the hexapole magnet and move into the low field region where scattering experiments are carried out, especially, when the magnetic field is close to zero and even a small field change can be quite abrupt for the atoms and can cause a change of state and hence a depolarization of the atomic beam. To prevent this depolarization, the atomic beam is guided adiabatically, using magnetic field carrying coils, from the hexapole magnet to the interaction region and to the Rabi detector, when required.

The polarization of the atomic beam in the strong magnetic field of the hexapole magnet should, theoretically, be 100% but actual polarization, inside the hexapole magnet, has been calculated by Hils et al. (1981) to be 70% (i.e. $P_A = 70\%$). In a weak magnetic field, however, the polarization is reduced by hyperfine structure coupling of "$I$" and "$J$" to a maximum theoretical value of

$$P_{max} = \frac{1}{(2I + 1)} \times 100\%, \tag{3.51}$$

where "$I$" is the nuclear spin of the atom. For the atoms of sodium and potassium $I = 3/2$, therefore, (3.51) gives a value of $P_{max} = 25\%$ for the polarization of the sodium and potassium atomic beams. Experimentally, the maximum polarization ($P_{max}$) achieved for alkali atoms is $\approx 21\%$, as measured in the present experiment using a combination of Rabi magnet and Langmuir-Taylor detector. Figure 3.45 shows such a profile, in the present study, for the potassium atomic beam and from this profile the polarization of the beam has been calculated.

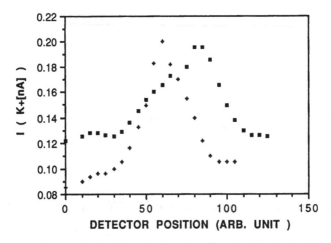

**Fig. 3.45** The atomic beam profile of potassium as measured by the Langmuir-Taylor detector (50 U along the $x$-axis is equal to 0.115 cm); (*plus sign*) refers to the undeflected beam when Rabi magnet is switched off; (*filled square*) refers to the deflected beam when Rabi magnet is switched on

### 3.3.5 The Rabi Magnet

An inhomogeneous magnetic field of a pattern which allows the spin analysis of atomic beams is provided by the Rabi magnet (Ramsey 1969). The magnet is made from soft iron which is magnetized by passing electric current through the wire wound coils. Figure 3.46 shows a Rabi magnet design with an advantage that its boundaries correspond to the equipotentials of the field produced by a system of two parallel wires (Ramsey 1969). Figure 3.47 shows the magnetic field around two wires and the equipotential curves. There is a uniform magnetic gradient over the beam height as shown in Fig. 3.46. The slit defining the beam has a width of 0.1 mm and a height of 8 mm. A permanent magnetic moment can be assigned to the valency electron, which it derives from an intrinsic angular momentum (spin), determined by a fixed quantum number $s = 1/2$. Therefore, sodium and potassium atoms, in ground state, when passing through the Rabi magnet suffer a deflecting force given by

$$F = \mu_H \left( \frac{dH}{dz} \right), \tag{3.52}$$

where $\mu_H = g_J m_J \mu_B$, in a strong magnetic field.

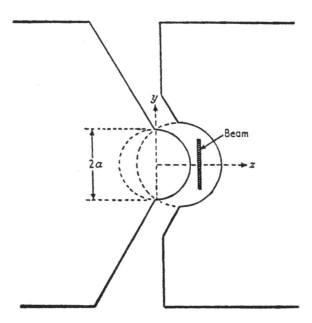

**Fig. 3.46** Cross-sectional view of the pole pieces of a Rabi magnet effectively producing two-wire magnetic field. The direction of the beam is perpendicular to the plane of the paper (after Ramsey 1969)

**Fig. 3.47** (——) shows the magnetic field around two current carrying wires; (- - -) shows the magnetic equipotential curves. Wires are assumed to be infinitely narrow. Outside the wires, however, the magnetic field is independent of the wire diameter

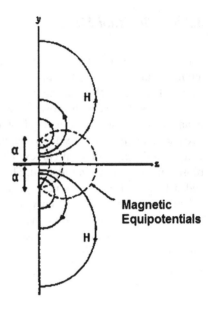

For sodium and potassium

$$m_J = \pm \frac{1}{2} \quad \text{and} \quad g_J = 2$$

$$\mu_H = \pm \mu_B \text{ and}$$

$$F = \pm \mu_B \left( \frac{dH}{dz} \right). \tag{3.53}$$

When the Rabi magnet is switched on, the beam profile splits into two nearly equal and well-resolved traces which allow the measurement of the polarization of the atomic beam (see Fig. 3.45). Polarization of the beam is given by

$$P = \frac{N_1 - N_2}{N_1 + N_2}, \tag{3.54}$$

where $N_1$ = number of atoms with spin up, and $N_2$ = number of atoms with spin down.

### 3.3.6 Langmuir-Taylor Detector and the Atomic Beam Density

Figure 3.48 shows a Langmuir-Taylor detector. This is a surface ionization detector which is made of a 3 cm long tungsten wire of 0.1 mm diameter, placed at the centre

**Fig. 3.48** Langmuir-Taylor detector

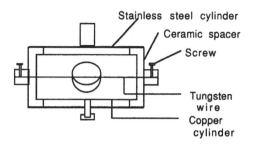

of a copper cylinder, and is oxidized by flashing it in air for a second. At the centre of the cylinder, a 1 cm diameter hole is made to allow the atomic beam to enter the detector. The detector is placed inside another cylinder and is isolated from the outer cylinder by ceramic spacers. A vertical drive held by bellows can move the detector up or down for adjustment purposes. When the atoms of the atomic beam hit the tungsten wire, which is heated by passing a suitable electric current through it, the atoms get ionized at the surface of the wire and since the wire is biased positively with respect to the cylinder these ions get collected by the copper cylinder. If $n^+$ and $n$ are, respectively, the number of ions and neutral atoms which re-evaporate from the surface of the hot wire then the ratio $n^+/n$ is given by

$$\frac{n^+}{n} = \exp\left(\frac{\varphi - I}{kT}\right),\tag{3.55}$$

where $I$ is the ionization potential of the atoms in eV, $T$ is the temperature of the wire, $\phi$ is the work function of the wire and $k$ is the Boltzman constant equal to $8.5 \times 10^{-4}$ eV/K. The condition for an atom striking the hot tungsten wire to get ionized is that

$$(\varphi - I) \geq 0.5\,\text{eV}.\tag{3.56}$$

Tungsten (free from thorium) has a work function of 4.48 eV, which can be increased by surface oxidation to 6 eV. Since the ionization potential for sodium $I_{\text{Na}} = 4.1$ eV and for potassium $I_{\text{K}} = 5.13$ eV, the oxidized tungsten wire can be used to detect both sodium and potassium atoms.

Figures 3.49 and 3.50 show, respectively, the beam profiles for sodium and potassium atoms. For potassium, for example, the density of the atomic beam at the detector ($n_{\text{det}}$), 112.5 cm away from the hexapole magnet, can be related to the atomic beam density ($n_{\text{ex}}$) at the exit of the hexapole magnet by the relationship

$$\frac{n_{\text{ex}}}{n_{\text{det}}} = \left[\frac{D_{\text{det}}}{D_{\text{ex}}}\right]^2,\tag{3.57}$$

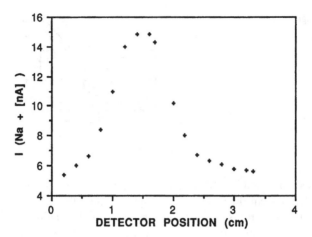

**Fig. 3.49** Sodium beam profile

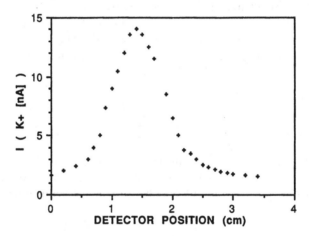

**Fig. 3.50** Potassium beam profile

where $D_{\text{det}}$ and $D_{\text{ex}}$ are, respectively, the diameters of the beam at the detector and the exit of the hexapole magnet. The number of detected atoms is given by

$$n_{\text{det}} = \frac{i_{\text{ion}}}{I_w D_w \underline{v} e}, \tag{3.58}$$

where $i_{\text{ion}}$ is the ion current, $I_w$ ($=1$ cm) is the length of the wire exposed to the atomic beam, $D_w$ is the wire diameter, $\underline{v}$ ($=(8kT/\pi M)^{1/2}$) is the average atomic velocity and $e$ is the electronic charge.

For sodium ($T = 660$K), $\underline{v} = 7.812 \times 10^4$ cm/s

and for potassium ($T = 573$K), $\underline{v} = 5.572 \times 10^4$ cm/s.

This gives the detected ion density for sodium

$$n_{\text{det}} = 1 \times 10^{14} \text{atoms/m}^3 \tag{3.59}$$

and for potassium

$$n_{\text{det}} = 1.5 \times 10^{14} \text{atoms/m}^3 \tag{3.60}$$

From (3.59) and (3.60), substituting the values in (3.57), we get for sodium

$$n_{\text{ex(Na)}} = 2.50 \times 10^{15} \text{atoms/m}^3 \tag{3.61}$$

and for potassium

$$n_{\text{ex(K)}} = 3.8 \times 10^{15} \text{atoms/m}^3 \tag{3.62}$$

The diameter of the atomic beam $(D_{\text{I.R}})$ at the interaction region, which is 38.5 cm away from the exit of the hexapole magnet, is

$$D_{\text{I.R}} = 0.41 \text{cm},$$

therefore, the density of the atomic beam $(n_{\text{I.R}})$ at the interaction region is given by

$$(n_{\text{I.R}}) = n_{\text{det}} \left[ \frac{D_{\text{det}}}{D_{\text{I.R}}} \right]^2.$$

For sodium

$$(n_{\text{I.R}}) = 6.0 \times 10^{14} \text{atoms/m}^3 \tag{3.63}$$

and for potassium

$$(n_{\text{I.R}}) = 9.0 \times 10^{14} \text{atoms/m}^3. \tag{3.64}$$

### 3.3.7   The Electron Beam Components

For this experiment, a low energy and high-current electron gun is required and Simpson and Kuyatt (1963) used such a gun with a multistage design. The principle of this gun is derived from the observation in the equipotential guns, such as diodes, that below a certain minimum voltage the electron beams of these guns cannot

**Fig. 3.51** *Dotted line* shows
an ideal space charge limited
beam profile required to
saturate a given space

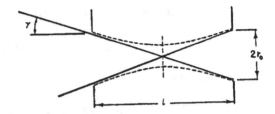

saturate a given space. Therefore, if an electron beam having energy $E$ passes
through a region of length ($l$) defined by two apertures of diameter $2r_0$ and if the
maximum convergence angle is $\gamma$ then there are two restrictions on the maximum
current which can pass through. The repulsive force between electrons when they
come very close, i.e. the space charge force, is the first restriction, and this force is
independent of $l$ and $r_0$ but depends on the ratio $l/r_0$. If an electron beam, in the
absence of any magnetic field, enters the space with a maximum angular spread of
$2\gamma$ such that the extreme rays would cross in the middle of the space, as shown in
Fig. 3.51, then the maximum current which can be put through this region is

$$I_{max} = 38.5E^{3/2}\left(\frac{2r_0}{l}\right)^2 = 38.5E^{3/2}\tan^2\gamma,$$

where $I_{max}$ is in $\mu$A and $E$ is in eV. An electron beam emerges in the shape of a disc
of diameter $2r_0/2.35$ and the electrons travel in the centre of the space parallel to the
axis of the apertures. The Helmholtz-Lagrange theorem, which gives the second
restriction to having a high-current electron beam, states that, for any two planes $Z_1$
and $Z_2$ separated by a non-absorbing optical path, such that plane $Z_2$ is the image of
plane $Z_1$, we have

$$n_1 dx_1 \sin\theta_1 = n_2 dx_2 \sin\theta_2.$$

In this equation, $n_1$ and $n_2$ are the indices of refraction, $\theta_1$ and $\theta_2$ are the angles of
convergence at points in the planes $Z_1$ and $Z_2$, and $dx_1$ and $dx_2$ are the differential
elements in the planes (Simpson 1967). Simpson and Kuyatt (1963) dealt with the
beam density limitations of such guns and obtained the following expression which
relates the cathode temperature $T$ (in K), the electron convergence angle $\theta_2$, and the
maximum convergence angle of the apertures $\gamma$ to the electron energy $E$

$$E \geq \left(\frac{211}{\pi}\right)\left(\frac{T}{11,600}\right)\left(\frac{\tan^2\gamma}{f(\gamma)\sin^2(\theta_2)}\right). \tag{3.65}$$

Figure 3.52 (from Simpson 1967) shows the regions that are accessible for
obtaining space charge saturation current, $I_{max}$. It is evident from Fig. 3.52 that
for $\gamma = 9.2°$, the maximum saturation current, $I_{max}$, cannot be obtained for anode
potential less than 220 V and that for 10 V, anode potential, the maximum current

that can be expected from such a gun is 25% of $I_{max}$, i.e. a current of approximately 20 μA, for a convergence angle of 17.85°. These difficulties can, however, be overcome, to a great extent, in a multistage electron gun by extracting the electrons at some higher energy to overcome the space charge problem, and then decelerating the electron beam to its final desired energy. In the electron gun, we have used, the cathode is biased negatively and the image of the cathode forms the plane of the smallest cross-section and this image of the filament is transferred through the various parts of the gun until it forms a reduced image at the interaction region.

Figure 3.53 shows the design of the three stage electron gun which we have used. The first stage, or the extraction stage, consists of a cathode (0), a grid (1) and an anode (2), which is made from tantalum and forms a Soa-type immersion lens having a property that the object and the image lie within the lens field. The lens, therefore, forms an image of the cathode at an intermediate position which is a function of $\gamma = v_g/v_a$ (Soa 1959). The grid bias influences greatly the position and the magnification of the cathode image and at zero or small negative bias the reduced image of the cathode appears close to the anode plane as shown by the graph in Fig. 3.54.

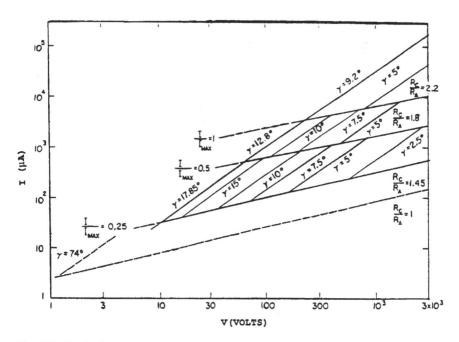

**Fig. 3.52** Graph of current versus voltage showing regions accessible to diode guns capable of obtaining space charge saturation current $I_{max}$ within a given space defined by convergence angle. Cases of less than full saturation, i.e. $0.5 I_{max}$ and $0.25 I_{max}$ currents are also shown. The usable region is to the right of the $\gamma$ = constant line and above $I/I_{max}$ = constant lines. On the right margin is given the cathode to anode radius $R_0/R_a$ of the optimum Pierce gun. The *dashed line* at the bottom is the zero anode–cathode spacing limit (after Simpson and Kuyatt 1963)

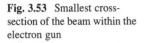

**Fig. 3.53** Smallest cross-section of the beam within the electron gun

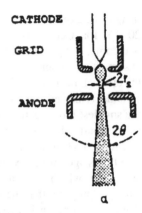

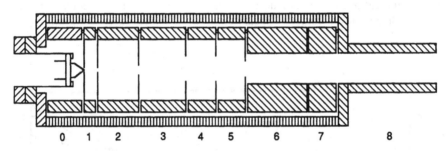

**Fig. 3.54** Cross-sectional view of the gun (to scale). (0) is the cathode; (1) is the grid; (2) is the anode; (3), (4) and (5) are the decelerating lenses; (6), (7) and (8) are the variable lenses

Three-aperture lenses (3, 4 and 5) which form the second stage, or the decelerating stage, allow a diminished image of the cathode to be projected into the interaction region. Although the first two stages of the gun form a good enough low-energy electron gun, yet a third has been added to avoid any electric or magnetic field near the interaction region. The third stage, therefore, consists of a variable ratio lens whose first element is kept at the same potential as the last element of the second stage and the last element of the third stage, i.e. the element nearest to the interaction region, is kept at earth potential. Since the cathode of the electron gun is kept at a negative potential and the last element of the gun is kept at earth potential, the net energy of the electron beam is given by the cathode bias after taking into account any contact potential.

The various parts of the electron gun are made from different materials which have different work functions with respect to the vacuum and, therefore, produce a net contact potential, $V_c$. Measurements have been made to derive the value of the contact potential for each experiment. In the case of potassium beam experiment, a photomultiplier in combination with an interference filter is used to detect the potassium line (404.4 nm, $5P_{1/2} - 4S_{1/2}$) radiation. The potential at the cathode is changed in steps of 1 V near the threshold for the excitation process and photons are

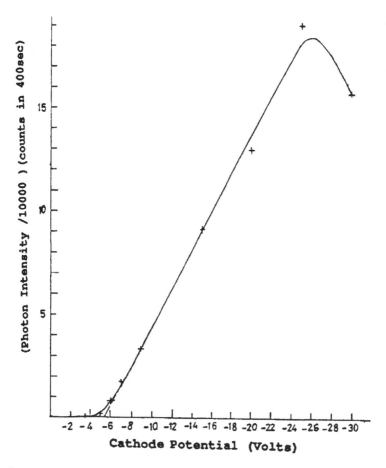

**Fig. 3.55** Determination of the contact potential

observed at each potential step. Figure 3.55 shows a graph of the photon count as a function of the cathode potential. The intersection, of the straight part of the curve with the zero intensity axis, gives the excitation potential of the $5P_{1/2}$ level which is known to be at 3.06 V and the difference between this and the observed value gives the contact potential, $V_c$. According to the graph in Fig. 3.54, the measured value for the excitation potential of the $5P_{3/2}$ state is found to be 5.5 V and, therefore, the contact potential is 2.44 V. For a 10 eV nominal energy value of the electron beam, for example, the true energy is 7.56 eV.

A Faraday cup, made from non-magnetic stainless steel, collects the electron beam. It consists of a cylinder of 0.8 cm diameter and 6.8 cm length and is enclosed in another cylinder of 1 cm diameter and 7.3 cm depth. A Teflon spacer is used to insulate the Faraday cup from the outer cylinder which is at earth potential. An electrometer (Keithley, model 610 CR) is connected to the Faraday cup to measure the electron beam current. The Faraday cup is covered with soot to avoid electron reflection and, therefore, improve the electron collection efficiency.

The various parts of the electron gun are either covered with soot or Aquadag to reduce electron and photon reflections.

### 3.3.8   Photon Detection System

In the electron-polarized-atom collision investigation, the induced fluorescence light is detected in the direction of the spin polarization, perpendicular to the electron and atomic beam directions. A pair of Helmholtz coils, where each coil of 20 turns has a diameter of 3 cm, are placed 1.5 cm apart and are used to control the magnetic field strength and direction in the interaction region. Further, by reversing the magnetic field direction in the Helmholtz coils, the spin polarization direction of the atoms can also be reversed. A Thurlby power supply PL370 is used to provide the Helmholtz coils the required stabilized constant current.

Figure 3.56 shows firmly held in the interaction region a pair of small Helmhotlz coils and a plano-convex lens of 75 mm focal length and 60 mm aperture to collect the emitted fluorescence radiation. The position of the lens is adjusted to give a magnified image of the interaction region at the entrance window of the photo-multiplier located 63 cm away outside the scattering chamber. The fluorescence

**Fig. 3.56** The arrangement of the apparatus at the interaction region

light collected from the interaction region is allowed to pass through a polarizer, a suitable interference filter for the radiation line of interest and a photomultiplier, which form a single unit and can be rotated manually. For the measurement of the linear polarization, the light collected by the lens is allowed to pass through the linear polarizer and a suitable interference filter, for the spectrum line of interest, before it is detected by the photomultiplier. For convenience of operation, the linear polarization measurement system consisting of the polarizer, the filter and the photomultiplier form a single unit which can be rotated manually. For the measurement of circular polarization of the fluorescence light, however, a suitable $\lambda/4$ plate is inserted in the path of the light before the linear polarizer in such a way that the fast axis is set parallel to the electron beam. To mark the fast and slow axes of the $\lambda/4$ plate, a Soleil-Babinet compensator (Halle Nachf) is used. In this experiment, different photomultipliers have been used for detecting the light of different wavelengths. For the sodium D-lines (588.9/9.5 nm), for example, a red-sensitive photomultiplier (EMI model D-624), which has a quantum efficiency of 11.8% for 580 nm wavelength at an anode potential of 1,600 V, and for the blue potassium line (404.4 nm) the EMI model 9883QA photomultiplier has been used. For potassium lines, the linear polarizer type HN-PB and the $\lambda/4$ plate, made of mica sandwiched between two glass plates (Dr. Steeg & Reuter, type B), have been used.

The photomultipliers, used in this experiment, have been shielded magnetically by using soft material (Conetic) and powered by Fluke high-voltage power supply, model 408 B. The anode of the photomultiplier has been biased positively while the cathode is kept at the earth potential. The output pulses from the photomultiplier are fed to an amplifier (Lecroy, model 612A). The amplified pulses are fed to a constant fraction discriminator (Ortec, model 473) to cut off the noise from the signal. The amplified low noise signal pulses are then counted by a rate-meter (Nuclear Enterprise, model 4607) and a linked scaler/timer (Nuclear Enterprise) system which recorded simultaneously the elapsed time and the number of detected photons.

The experimental technique for this experiment is described in Sect. 4.3, the results and discussion are given in Sect. 5.3 and the concluding remarks are mentioned in Chap. 6.

## 3.4 Apparatus for the Study of Excitation of Calcium

### 3.4.1 The Vacuum System

The high-vacuum system consists of a chamber in the form of a cylinder having inner diameter of 350 mm and a length of 600 mm and is made of non-magnetic stainless steel. The vacuum chamber (see Fig. 3.57) has ports to connect the apparatus such as high-vacuum pump and introduce components like photon detector assembly and Faraday cup. Copper gaskets and Viton o-rings are used to vacuum seal the flanges. The chamber is pumped by a turbo-molecular pump

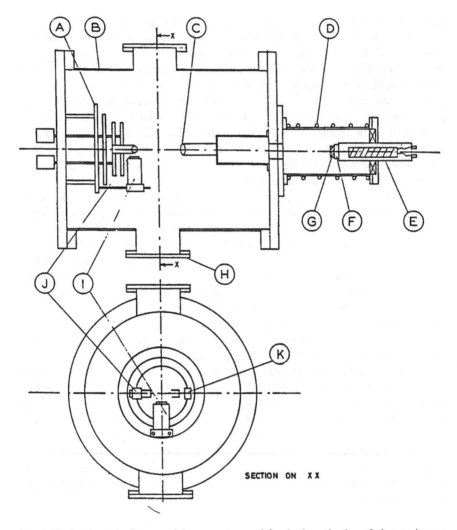

**Fig. 3.57** A schematic diagram of the apparatus used for the investigation of electron impact excitation of alkaline-earth metal atoms

(Balzers, TPU 510) having a pumping speed of 500 ls$^{-1}$ for air. A two stage rotary pump (Edwards, ED M20 A) is used as a backing pump for the system.

### 3.4.2  The Atomic Beam Source

A thermal beam of calcium atoms is formed inside the vacuum chamber by effusive flow from an oven (Figs. 3.58 and 3.59) heated upto ∼700 °C. The oven is made from stainless steel and consists of two parts. The lower part of the oven serves as a

**Fig. 3.58** A cross-sectional
view of the oven

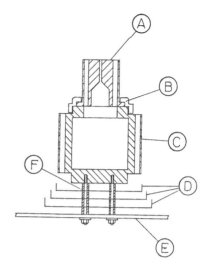

**Fig. 3.59** A photograph of
the oven

reservoir for calcium metal and holds approximately 40 g of calcium. The upper
part consists of a nozzle, which has a channel of 8 mm length and 0.8 mm inner
diameter, through which calcium atoms are allowed to escape into the surrounding
vacuum. The oven is heated by two independent sets of coils made from tantalum

wire of 0.25 mm diameter. The lower part of the oven containing calcium has 24 holes for inserting heating coils. A separate heating coil is used to heat the nozzle. A special fourfold wiring has been used for the heating coils to avoid having any magnetic field due to the heating current. The temperature of the nozzle is always kept higher than the lower part of the oven to avoid clogging of the nozzle channel. Three cylindrical stainless steel heat shields with top and bottom made from tantalum surround the oven and provide a good thermal insulation. The heat shields are insulated from each other and from the base of the oven by thin ceramic spacers and held in position by the mounting rods. Small diameter apertures are used to collimate the atomic beam.

### 3.4.3   Atomic Beam Density

The oven is designed to provide an intense and well-collimated beam of calcium atoms. In order to obtain an estimate for the atomic beam density (see also Sect. 3.3.3.2) we start from the number $dQ$ of atoms which emerge per second from a source slit travelling into a solid angle $dw$ at an angle $\theta$ with respect to the normal to the slit plane. The value of $dQ$ is given by Ramsey (1956) as:

$$dQ = \frac{dw}{4\pi} n\bar{u}(\cos\theta)A_s, \qquad (3.66)$$

where $n$ is the number of atoms per unit volume, $\bar{u}$ is the mean atomic velocity and $A_s$ is the area of the source slit. Equation (3.66) implies the cosine law of molecular effusion from a thin-walled aperture. However, if a canal-like aperture of appreciable length is used instead, the angular spread of the emergent beam can be reduced considerably (Clausing 1929). The beam formation by long tubes has been given by Ramsey (1956), Giordmaine and Wang (1960) and Lucas (1973). They have shown that the peak intensity and the collimation of the beam are essentially determined by the interatomic collisions in the tube. Depending on the ratio of the mean free path length $\lambda$ in the tube to the radius $r$ and length $l$ of the tube, different results for the beam intensity are obtained. At pressures such that

$\lambda \geq 1, \lambda \gg r$, i.e. when atom-atom collisions do not occur in the tube, we get optimum intensity and collimation (Lucas 1973). The mean free path $\lambda$ in cm is related to the pressure $P$ in torr through the relation (Ramsey 1956)

$$\lambda = 7.321 \times 10^{-20} \frac{T}{P\sigma}, \qquad (3.67)$$

where $T$ is the temperature in Kelvin. $\sigma$ is the atomic collision cross-section in $cm^2$ and is equal to $\pi\delta^2$ where $\delta$ is the diameter of the atom. For $T = 800$ K, the vapour pressure $P = 2 \times 10^{-2}$ torr and atomic radius of calcium atom being 1.69 Å, the mean free path

$$\lambda = 0.82 \, \text{cm.} \tag{3.68}$$

Ramsey (1956) also showed that for a channel-like aperture, the total number of atoms of mass $M$ effusing per second is given by

$$Q = \frac{l}{R} \frac{n\bar{u}A_s}{4}, \tag{3.69}$$

where

$$\frac{l}{R} = \frac{8r}{3l}, \quad l >> r,$$

$$\bar{u} = 1.13\sqrt{\frac{2KT}{M}} = 6.4 \times 10^4 \text{cm/s} \; (K \text{ is the Boltzmann costant})$$

is the mean atomic velocity inside the source, $n = 7 \times 10^{14}$ atoms/cm$^3$ is the number of atoms per unit volume at $2 \times 10^{-2}$ torr pressure and $A_s = \pi r^2 = 785 \times 10^{-5}$ cm$^2$ is the area of the source slit. Substituting these values in (3.69) we obtain

$$Q = 13.8 \times 10^{15} \text{atoms/s} \tag{3.70}$$

The forward peak intensity at a distance $l$ from the source is given by Ramsey (1956) as

$$I = \frac{1}{4\pi} \frac{A_d}{l^2} n\bar{u}A_s, \tag{3.71}$$

where $A_d$ is the area of a detector placed at a distance $l$ from the source.

Substituting $A_d = 1$ cm$^2$ and $l = 3$ cm, which is the distance from the interaction region to the beam source, the intensity (or flux) in the interaction region is

$$I = 1.4 \times 10^{15} \text{atoms/cm}^2\text{s.} \tag{3.72}$$

Hence, the beam density of calcium atoms ($N_{Ca}$) at the interaction region is given by

$$N_{Ca} = \frac{I}{\bar{u}} = 2 \times 10^{10} \text{atoms/cm}^3. \tag{3.73}$$

For better experimental results, the oven is heated to 920 K to obtain a higher beam density

$$N'_{Ca} = \frac{I}{\bar{u}} = 1.4 \times 10^{11} \text{atoms/cm}^3. \tag{3.74}$$

### 3.4.4    The Electron Gun

The electron gun used in these investigations is a simple, high-performance electrostatically focused electron gun which provides a useful electron beam over an energy range from a few electron volts to ~500 eV. The gun has been constructed according to the description given by Erdman and Zipf (1982) and is schematically illustrated in Fig. 3.60. The gun consists of a filament F mounted on a block A followed by a three-element lens system (B, C and D) and an aperture E at the end kept at ground potential. All elements of the gun are cylindrical in shape and are made from stainless steel. Ruby balls of 3 mm diameter are used to separate and insulate the various elements of the gun.

The required energy of the electron beam is set up by the difference of the voltage on the cathode of the electron gun and the earth potential. The voltages applied to the various elements of the gun are then optimized by monitoring the electron beam at the Faraday cup which has been mounted on a separate turntable. A typical result is shown in Fig. 3.61 for electron energy of 40 eV. The full width at half maximum (FWHM) of this curve is ~8°.

## 3.5    Apparatus for the Study of Excitation of Strontium

### 3.5.1    The Vacuum Chamber

The vacuum chamber is the same as described in Sect. 3.4.1.

### 3.5.2    The Atomic Beam Source

The atomic beam source is same as described in Sect. 3.4.2.

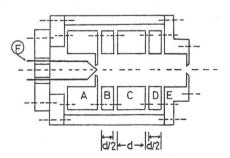

**Fig. 3.60** A schematic diagram of the electron gun assembly. The aperture plates in the lens elements A and B are each 0.1 mm thick

**Fig. 3.61** Angular
distribution of the electron
beam as measured by the
Faraday cup at an incident
electron energy of 40 eV.
FWHM of the beam is
$\sim8^{\circ}$ and maximum resolution
of the Faraday cup is $\sim4^{\circ}$

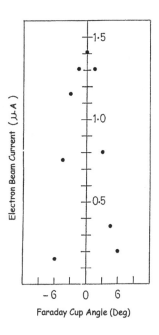

### 3.5.3   The Atomic Beam Density

The oven described in Sect. 3.4.3 has been used to get a beam of strontium atoms.
About 25 g of strontium metal is introduced in the main body of the oven and is
heated to a temperature of 550 °C. The nozzle is always kept at a temperature higher
than 550 °C. A beam density of $\sim1 \times 10^{10}$ atoms/cm$^3$ and a beam width of 10 mm
at the interaction region is obtained.

### 3.5.4   The Electron Gun

The electron gun used in this study is the same as described in Sect. 3.4.4. An
electron beam having a current of 1 μA and a diameter of 8 mm at the interaction
region is easily obtained at 45 eV energy.

### 3.5.5   The Electron Energy Analyzer

The electron energy analyzer used in these investigations is a 127° cylindrical
analyzer similar to the one described in Sect. 3.1.5. The analyzer is fixed to one
of the platforms of a triple-turntable so that the electron scattering angle can be set

from outside the vacuum chamber to be within $\pm 3°$ and measured with respect to the direction of the electron beam. The angular range is limited on the low side to about 30° by the spread of the electron beam and on the high side to about 150°, when the electron analyzer hits the electron gun. To be detected, the scattered electrons have to pass through a cone with an acceptance angle of 0.015 sr. The electrons after passing through the cone are focused and directed onto the entrance slit of an electrostatic lens system and then analyzed for energy. The electrons of selected energy are collected as they come out of the exit slit of the analyzer by a channel electron multiplier (Mullard, B318L).

## 3.6   Apparatus for the Study of Excitation of Helium

Figure 3.62 shows a photograph of the apparatus which consists of two stainless steel cylindrical chambers 53 and 43 cm long sealed together to form one long cylindrical chamber suspended horizontally. Both ends of this long cylindrical chamber are vacuum sealed with 57 cm diameter flanges using aluminium and copper gaskets. There are several ports for connecting other components. The chamber is evacuated to high vacuum using two diffusion pumps (Diffstak, Edwards model 160/700P) and a rotary oil backing pump. The vacuum chambers can be isolated from the pumping system by the use of butterfly valves.

The main components of the apparatus are a 127° cylindrical electrostatic monochromator, a 127° electrostatic electron analyzer, a gas nozzle, an electron gun and a photon detector system. Figure 3.63 shows schematically the components of the apparatus inside the vacuum chamber. A Pierce-type electron gun followed by an aperture lens focuses the electrons extracted from the tungsten filament onto the entrance slit of the 127° monochromator. The energy-selected electrons are allowed to pass through a 1 mm diameter collimating aperture before accelerating

**Fig. 3.62**  A photograph of the apparatus used for the electron impact excitation of helium atoms

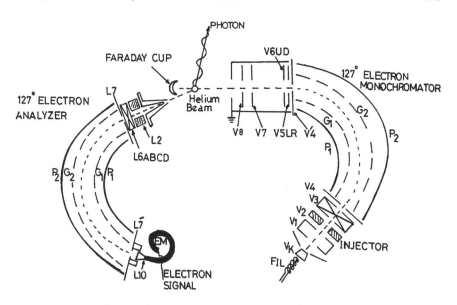

**Fig. 3.63** A schematic diagram of the components of apparatus inside the vacuum chambers

and focusing the electrons on the gas beam by using a two-aperture and a three-aperture lens system. The gas beam is produced by effusing the gas through a capillary tube mounted in a non-magnetic stainless steel cylindrical tube 25 mm long having a diameter of 1.0 mm. The interaction region is about 3 mm above the gas tube.

Electrons scattered in a particular direction and moving in a field-free region enter a cone of acceptance angle $7 \times 10^{-3}$ sr. and are then accelerated and focused onto the entrance slit of the electron analyzer. A channel electron multiplier is used to detect the electrons transmitted by the analyzer. The electron analyzer is mounted on a turntable which can be rotated from outside the vacuum chamber through angles ranging from $-90°$ to $+90°$ with respect to the electron beam direction.

When helium is used as target gas, the photons emitted by the excited helium atoms are detected in a direction perpendicular to the scattering plane by a photon detector system mounted outside the vacuum chamber. A fixed lens of a large collection solid angle (0.45 sr.) inside the vacuum chamber focuses the radiation from the interaction region onto the photon detector. The photon detector system consists of a linear polarizer followed by an interference filter and a photomultiplier tube (EMI 9883QA). A quarter wave-plate is placed in front of the linear polarizer to analyze circularly polarized light. The detector system can be rotated with respect to the quarter wave-plate axis and the beam axis.

A schematic diagram of the timing electronics used in the measurements is shown in Fig. 3.64. The detected electron and photon pulses are fed into the amplifiers (LRS 123 dual linear amplifier for electrons and LRS 234 linear amplifier

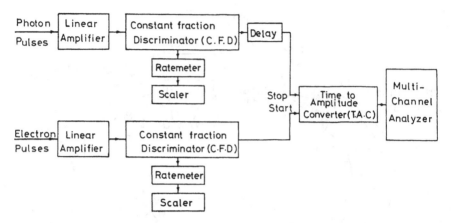

**Fig. 3.64** A schematic diagram of the timing electronics for the measurements

for photons). The amplified pulses are fed into the constant fraction discriminators (CFD) (ORTEC 473 for electrons and ORTEC 453 for photons). The electron timing pulses from the CFD provide the start pulse for the time-to-amplitude converter (TAC) (ORTEC 467) and the photon timing pulses after suitable delay are used to stop the TAC. Output signal pulse from the TAC is fed to a MCA (ORTEC 7100) in the PHA mode. Inelastically scattered electrons and emitted photons from the same event arrive with a definite time correlation forming a peak in the MCA spectrum, over and above the background.

Sections 4.12 and 5.5 and Chap. 6 describe the experimental technique, results and discussion and concluding remarks, respectively, for the study of 3 $^1$P excited state of helium and Sects. 4.13 and 5.6 and Chap. 6 give the experimental technique, results and discussion and concluding remarks, respectively, for the three $^3$P excited state of helium.

# Chapter 4
# Experimental Techniques

**Abstract** Experimental techniques for the investigation of electron impact excitation of helium, calcium and strontium atoms and for the measurement of the polarization parameters of the induced fluorescent radiation in electron-polarized-atom collision process are described. Also the electron-ion and X-ray-ion coincidence techniques for the investigation of electron impact ionization of atoms and molecules are discussed.

**Keywords** Excitation · Experimental techniques · Ionization · Polarization

## 4.1 Electron-Ion Coincidence Technique for the Investigation of Multiple Ionization of Atomic Gases

Figure 4.1 shows schematically the experimental setup for the electron-ion coincidence experiment to study the electron-atom collision process and to measure doubly differential cross sections (DDCS) and partial doubly differential cross sections [PDDCS or DDCS($n^+$)]. A focused beam of energetic electrons (Sect. 3.1.3) is allowed to interact with a beam of atoms emitted by a capillary array (Sect. 3.1.4). The Faraday cup (Sect. 3.1.3) collects the electron beam efficiently. The diameters of the beams of atoms and electrons are each about 2 mm. The interacting electrons are monoenergetic, while the atoms have a thermal energy distribution. Electrons ejected at 90° to the two beam directions and in a solid angle $\Delta\Omega$ are accepted by a 30° parallel plate electrostatic analyzer (Sect. 3.1.5), and the energy analyzed electrons are detected by a channeltron. To select ejected electrons having an energy $E$ (eV), a deflecting voltage of 0.6 EeV has to be applied to the upper plate of the analyzer while the lower plate is at ground potential. Electric potentials at the channeltron are varied so that the electrons incident at the

A. Chaudhry and H. Kleinpoppen, *Analysis of Excitation and Ionization of Atoms and Molecules by Electron Impact*, Springer Series on Atomic, Optical, and Plasma Physics 60, DOI 10.1007/978-1-4419-6947-7_4, © Springer Science+Business Media, LLC 2011

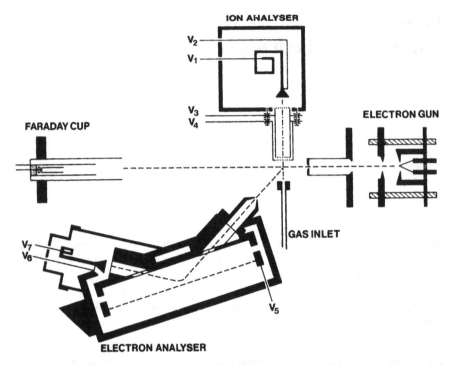

**Fig. 4.1** Experimental setup for the electron-ion coincidence spectroscopy. Inelastically scattered electrons from the region of the electron atom collision are energy analyzed and detected in coincidence with the produced ions, analyzed and detected by the ion analyzer

**Fig. 4.2** Output pulse from the electron analyzer. $\Delta t = 0.01$ μs and $\Delta V = 40$ mV

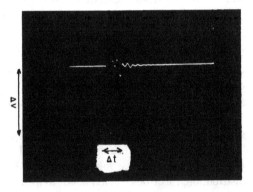

channeltron always have an energy of 200 eV for maximum detection efficiency (Mullard data book). Figure 4.2 shows an output pulse from the electron analyzer while the lower plate is at ground potential.

Ions produced by the collision process are extracted by a small electric field between 15 and 25 V/cm and are directed into a time of flight (TOF) type of ion analyzer (Sect. 3.1.6). Inside the ion analyzer, ions are further accelerated before they are allowed to drift in a 35-mm long field-free region inside the drift tube. At the end of the drift region, the ions are accelerated by a suitable high voltage at the cone of the channeltron for high efficiency of detection (Ravon 1982). Potential, $V_7$, at the closed end of the channeltron can be varied to keep the amplification of the channeltron constant throughout the experiment. Figure 4.3 shows an output pulse from the ion analyzer.

Figure 4.4 shows the electronic setup for measuring coincidences between the ejected electrons and the ions produced in the collision process. Pulses from the electron analyzer are amplified and fed to a timing discriminator, which supplies a prompt pulse suitable for the "start" of a time-to-amplitude converter (TAC). Also

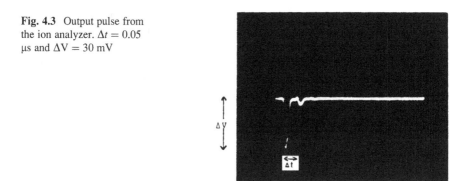

**Fig. 4.3** Output pulse from the ion analyzer. $\Delta t = 0.05$ μs and $\Delta V = 30$ mV

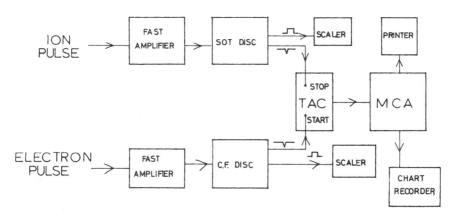

**Fig. 4.4** Electronic circuit for electron-ion coincidence spectroscopy

**Fig. 4.5** Time of flight
(TOF) spectrum for argon
ions

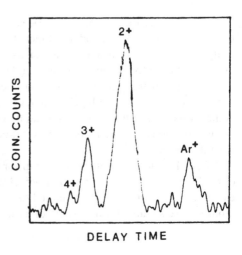

amplified pulses from the ion analyzer are fed to the input of a snap-off timing discriminator. The negative output of the timing discriminator is used to stop the TAC. The time delay between the pulses from an ion and ejected electron gives information about the charge state of the detected ion. Figure 4.5 shows such a time spectrum for argon. Peaks for charge state 1+ to 4+ can be seen in the figure. Delay time $(t_n)$ for the arrival of an ion with charge $n+$ is inversely proportional (3.6) to the square root of $n$, in any such spectrum.

## 4.2   X-Ray-Ion Coincidence Technique

Figure 4.6 shows schematically the experimental setup used for the X-ray-ion coincidence experiment. A focused beam of accelerated electrons from an electron gun (Sect. 3.1.3) is crossed with a beam of thermal atoms emitted by a capillary array (Sect. 3.1.4). Inner-shell ionization of atoms results in the emission of Characteristic X-rays (Sect. 2.1.7), which are detected by a hyper-pure germanium (HPGe) detector (Sect. 3.1.9).

Ions produced as a result of the collision process are extracted by a small field of about 25 V/cm and are directed into a TOF-type analyzer (Sect. 3.1.6).

Figure 4.7 shows the electronic setup for measuring coincidences between ions and emitted X-rays. X-ray pulses from the HPGe are fed to the timing filter amplifier (TFA) (Ortec 474). A fast negative pulse from the TFA is used as a start pulse for the TAC (Ortec 467). A stop pulse for the TAC comes from the ion, which is detected by the ion analyzer. The output pulse from the TAC is fed to a multi-channel analyzer (MCA) in pulse height analyzer (PHA) mode. A delay time spectrum for X-ray-ion coincidences is built up in the MCA. The time delay between the detected X-ray and the ion gives information about the charge state of the product ion.

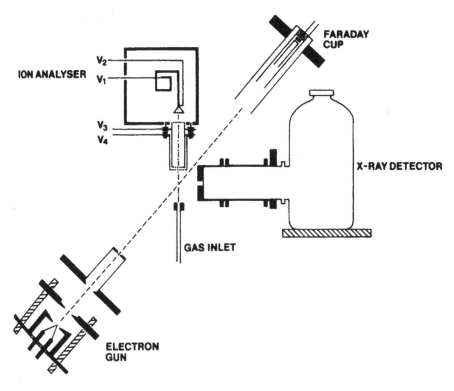

**Fig. 4.6** Experimental setup for X-ray-ion coincidence spectroscopy. X-rays from the region of the electron-atom collision are detected by the liquid nitrogen cooled hyper-pure germanium (HPGe) detector in coincidence with the produced ions, analyzed and detected by the ion analyzer

## 4.3 Crystal X-Ray Spectrometer

Figure 4.8 shows the crystal X-ray spectrometer setup for the X-ray spectroscopy of $K\alpha_1$ and $K\alpha_2$ lines of $^{54}$Mn from the decay of a $^{55}$Fe radioactive source. The X-ray source is placed in front of the Soller collimator (Sect. 3.1.12.3). The collimated beam of X-rays is incident on a NaCl(100) or LiF(200) crystal mounted on a plastic base. The crystal can be rotated through an angle of 0.01° steps with the help of a stepping motor and a gear system. The X-rays reflected according to the Bragg law are detected by a constant gas flow type proportional counter (Sect. 3.1.12.5). An argon-methane mixture (90% Ar and 10% methane) flows through the proportional counter constantly at a rate of $\sim$15 ml/min.

Figure 4.9 shows the electronic circuit for controlling the stepping motor and the MCA with the help of a BBC microcomputer. When a simple program in Basic is run in the computer, the MCA is switched to the collect mode and then with each step of the stepping motor the next channel of the MCA (in MCS mode) is made to collect the counts from the detector. At the end of the scanning period, the MCA is

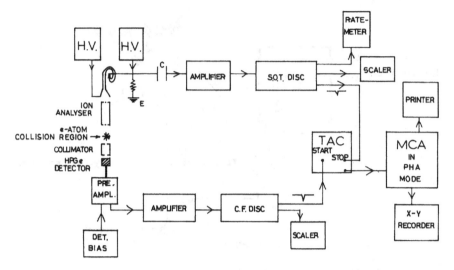

**Fig. 4.7** Electronic circuit for the X-ray-ion coincidence spectroscopy. Amplified pulse from the X-ray detector is fed to the start of time-to-amplitude converter (TAC) and an amplified pulse from the ion analyzer is fed to the stop of the TAC. The output pulse from the TAC is fed to the multi-channel analyzer (MCA) in pulse height analysis (PHA) mode. A time of flight type spectrum of the detected ions is obtained

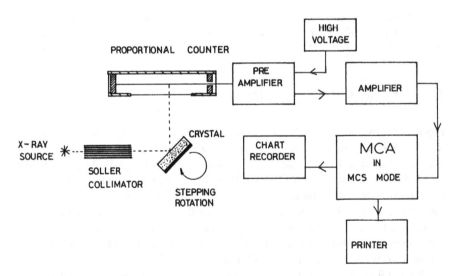

**Fig. 4.8** Experimental setup for the crystal X-ray spectroscopy. X-rays are collimated by the Soller collimator and then are made incident on the rotating crystal. The scattered X-rays are detected by a proportional counter. The electronic pulse from the proportional counter is amplified and fed to a multi-channel analyzer (MCA) in MCS mode (see text). A pulse height spectrum is formed in the MCA

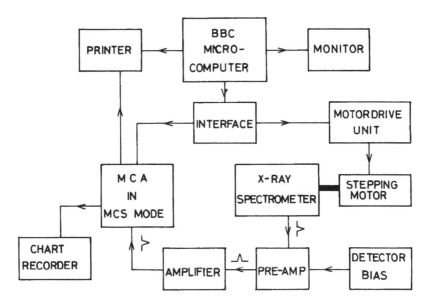

**Fig. 4.9**  Electronic circuit for the crystal X-ray spectrometer. The crystal in the X-ray spectrometer is rotated by a stepping motor which is controlled by a BBC micro-computer (see text)

switched out of the collect mode and the stepping motor brings the crystal to its original angular position, ready for the next scan. The time for each channel and the number of scans can be controlled to build up a good spectrum in the MCA.

## 4.4  Experimental Technique for the Investigation of the Ionization of the Molecular Gases

Figure 4.10 shows schematically the experimental setup for the study of the ionization of molecular gases by electron impact using the electron-ion coincidence technique. A focused beam of energetic electrons (Sect. 3.1.3) is allowed to interact with a beam of molecules emitted by a capillary array (Sect. 3.1.4). The Faraday cup (Sect. 3.1.3) collects the electron beam efficiently. The diameters of the beams of molecules and electrons are each about 2 mm. The interacting electrons are monoenergetic, while the molecules have a thermal energy distribution. Electrons ejected at $90°$ to the two beam directions and in a solid angle $\Delta\Omega$ are accepted by a $30°$ parallel plate electrostatic analyzer (Sect. 3.1.5), and the energy analyzed electrons are detected by a channeltron. To select the ejected electrons having energy $E$ (eV), a deflecting voltage of $0.6\,E$ eV has to be applied to the upper plate of the analyzer while the lower plate is at the ground potential. Electric potentials at the channeltron are varied so that the electrons incident at the channeltron always have an energy of 200 eV for maximum detection efficiency (Mullard

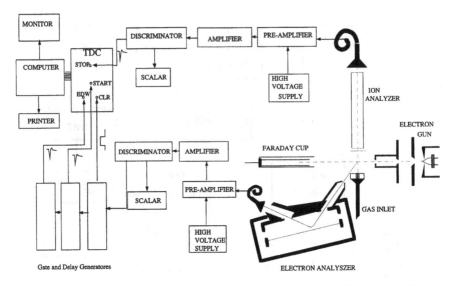

**Fig. 4.10** Electronic circuit for the electron-ion spectroscopy. Inelastically scattered electron from the electron molecule collision region is energy analyzed and detected by the electron analyzer. The detected electron pulse is amplified and fed to the start of the time-to-digital converter (TDC) and an amplified pulse from the ion analyzer is fed to the stop of the TDC. The data from the TDC are fed to a computer, which shows on the monitor a time of flight spectrum of the detected ions

data book). Ions produced by the collision process are extracted by a small electric field between 15 and 25 V/cm and are directed into a TOF type of ion analyzer (Sect. 3.1.6). Inside the ion analyzer, ions are further accelerated before they are allowed to drift in a 35-mm long field-free region inside the drift tube. At the end of the drift region, the ions are accelerated by a suitable high voltage at the cone of the channeltron for high efficiency of detection (Ravon, 1982). Potential, $V_7$, at the closed end of the channeltron can be varied to keep the amplification of the channeltron constant throughout the experiment.

The amplified pulse from the channeltron of the electron analyzer, after passing through a constant fraction discriminator, is fed to the start of a gate-and-delay generator, and from the gate-and-delay generator a fast pulse is fed to the start of the TDC. An amplified pulse from the channeltron of the ion analyzer is allowed to pass through the discriminator before feeding it to the stop terminal of the TDC. The output of the TDC is fed to the computer for processing the data. A TOF spectrum of the ions detected in coincidence with the electron can be seen in the computer.

Without any background, the TOF spectrum can be approximated to a Gaussian distribution and the amplitude, or the counts at each channel of the spectrum, is given by

$$y(x) = Ae^{\frac{(x-c)^2}{2w^2}},\tag{4.1}$$

where $x$ is the channel number, $A$ is the amplitude, $c$ is the central position of the peak and $w$ is the peak width. The area under the peak is (Kokta 1973)

$$N_c = 1.064AR_0, \tag{4.2}$$

where $R_0$ is the resolution of the detector, which is related to the peak's full width at half maximum (FWHM) by

$$R_0 = 2.355w. \tag{4.3}$$

The TOF spectrum, however, has always some background due to random coincidences, electron scattering, residual gas in the vacuum chamber, high counting rate, etc. To find the number of true coincidences, therefore, the background counts have to be subtracted from the number of coincidences under each peak. A typical spectrum is shown in Fig. 4.11. True coincidences can be found, in a spectrum where the background is not flat, using the total peak area (TPA) method given by Kokta (1973). If $N_1$ and $N_2$ are the number of coincidences under peak 1 and peak 2, respectively, then (see Fig. 4.10)

$$N_1 = \sum_{i=X_3}^{i=X_2} N_i - \frac{(N_{X_2} + N_{X_3})(X_3 - X_2 + 1)}{2} \tag{4.4}$$

and

$$N_2 = \sum_{i=X_4}^{i=X_5} N_i - \frac{(N_{X_4} + N_{X_5})(X_5 - X_4 + 1)}{2}, \tag{4.5}$$

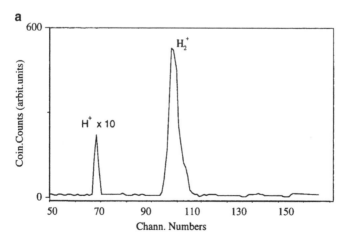

**Fig. 4.11**  Time of flight (TOF) spectrum of hydrogen molecule. Peaks for $H_2^+$ and $H^+$ ions can be seen in the spectrum

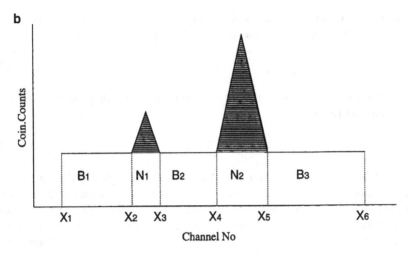

**Fig. 4.12** Data acquisition for analysis

where $N_i$ is the number of counts in channel $i$ (within the peak area) and $X_2, X_3, X_4$ and $X_5$ are the channel numbers as shown in Fig. 4.12. The background is taken from an area of the spectrum where there is no peak. Let $B_1, B_2$ and $B_3$ be the number of random coincidences between $(X_2 - X_1)$, $(X_4 - X_3)$ and $(X_6 - X_5)$, respectively, then the number of random coincidences $B$ per channel in the spectrum is given by

$$B = \frac{B_1 + B_2 + B_3}{(X_2 - X_1 + 1) + (X_4 - X_3 + 1) + (X_6 - X_5 + 1)} \tag{4.6}$$

and the number of true coincidences $N_c(n)$ under each peak is

$$N_c(1) = N_1 - B(X_3 - X_2 + 1) \tag{4.7}$$

and

$$N_c(2) = N_2 - B(X_5 - X_4 + 1). \tag{4.8}$$

The error $\Delta N_c(n)$ in the true coincidences is given by

$$\Delta N_c(1) = \sqrt{N_1 + \left[ \frac{X_3 - X_2 + 1}{(X_2 - X_1 + 1) + (X_4 - X_3 + 1) + (X_6 - X_5 + 1)} \right]^2 (B_1 + B_2 + B_3)} \tag{4.9}$$

$$\Delta N_c(2) = \sqrt{N_2 + \left[ \frac{X_5 - X_4 + 1}{(X_2 - X_1 + 1) + (X_4 - X_3 + 1) + (X_6 - X_5 + 1)} \right]^2 (B_1 + B_2 + B_3)}. \tag{4.10}$$

By normalizing to the total number of detected ions ($N_i$), the effects of the fluctuations in the target density, electron beam current and the data collection time are eliminated from the values of the true coincidences $N_c(n+)$ and the statistical error $\Delta N_c(n+)$. The values $N_c(n+)/N_i$ and $\Delta N_c(n+)/N_i$ are then used to find the partial double differential cross section (PDDCS or DDCS($n+$) or $d^2\sigma(n)/dEd\Omega$) for the production of ($n+$) ion by the relation (Hippler et al. 1984)

$$\text{PDDCS} = \text{DDCS}(n+) = \frac{d^2\sigma(n+)}{dEd\Omega} = \frac{N_c(n+)\sigma_i}{N_i\Delta E\Delta\Omega\varepsilon_\delta} \pm \frac{\Delta N_c(n+)\sigma_i}{N_i\Delta E\Delta\Omega\varepsilon_\delta}, \tag{4.11}$$

where $\sigma_i$ is the total ionization cross section for the target and is taken from the literature, $\varepsilon_D$ is the efficiency of the electron detection system, $\Delta E$ and $\Delta\Omega$ are the energy bandwidth and the solid angle, respectively. No attempt is made to determine the factor $\Delta E\Delta\Omega\varepsilon_\delta$. However, we have tried to keep the factor $\Delta E\Delta\Omega\varepsilon_\delta$ constant throughout the experiment. Therefore, only relative values of PDDCS are found.

The double differential cross section (DDCS) for the ionization of a molecule is obtained by summing the DDCS($n+$) over all the produced ions, i.e.,

$$\text{DDCS} = \sum_n \text{DDCS}(n+). \tag{4.12}$$

This value of the DDCS can then be normalized to an experimental value from the literature.

The results and discussion for this experiment are described in Sect. 5.2 and the conclusion in Chap. 6.

## 4.5 Experimental Technique for the Measurement of the Polarization Parameters of the Induced Fluorescent Radiation in Electron-Polarized Atom Collision Process

Figure 4.13 shows schematically the experimental arrangement for the measurement of the polarization of the induced fluorescence radiation from the electron-polarized atom collision process for sodium and potassium atoms. The atomic beams of sodium and potassium, produced by heating pure sodium and potassium metals in the oven, are spin polarized by passing through a hexapole magnet before allowing them to collide with the electron beam in the interaction region. The induced fluorescence light from the interaction region is collected by a lens and analyzed for the polarization parameters before detection by a cooled photomultiplier. The electronic pulses from the photomultiplier are amplified and counted for each experiment. The spin polarization of the atomic beam can be measured by the Rabi-magnet and Langmuir-Taylor detector system shown in Fig. 4.13. The magnetic coil, shown next to the hexapole magnet, can be used to flip the polarization of the atomic beam.

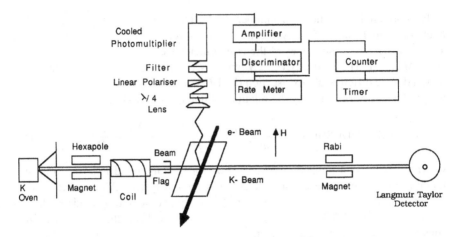

**Fig. 4.13** A schematic diagram of the experimental setup for the measurement of polarization parameters for the fluorescence radiation emitted by electron impact excitation of spin-polarized sodium and potassium atoms. An atomic beam obtained by heating alkali metals in an oven is made partially spin-polarized by allowing it to pass through a hexapole magnet. The spin-polarized atomic beam is then allowed to cross a beam of low-energy monoenergetic electrons. The excited alkali metal atoms emit fluorescence radiation which is analyzed for its polarization and then detected by a cooled photomultiplier. The photomultiplier pulses are amplified and counted

## 4.6   The Polarization Parameters of the Fluorescent Radiation

Figure 4.14 shows a general arrangement of a retarder, of retardance $\Delta$, and a polarizer for the measurement of polarization parameters. Born and Wolf (1970) have given a method for the measurement of polarization parameters for such an arrangement. If $\alpha$ is the angle of the polarizer transmission axis with respect to the $y$-axis (or the electron beam direction) and $\beta$ is the angle of the fast axis of the $\lambda/4$ retarder with respect to the $y$-axis, then the intensity of the fluorescent light is given by $I(\alpha, \beta)$. The intensities of the light transmitted through the retarder-polarizer system can be written as

$$I_1 = I(0°, 0°) + I(90°, 90) \tag{4.13a}$$

$$I_2 = I(45°, 45°) + I(135°, 135) \tag{4.13b}$$

$$I_3 = I(45°, 0°) + I(135°, 0). \tag{4.13c}$$

The linear polarization Stokes parameter $P_1$ is given by

$$P_1 = \frac{I(0°, 0°) - I(90°, 90°)}{I_1} \tag{4.14}$$

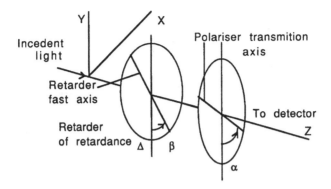

**Fig. 4.14** Optical set up for the measurement of polarization parameters of the fluorescence radiation emitted by the excited atoms of alkali metals

and the linear polarization Stokes parameter $P_2$ is given by

$$P_2 = \frac{I(45°, 45°) - I(135°, 135°)}{I_2}. \tag{4.15}$$

The circular polarization Stokes parameter $P_3$ is given by

$$P_3 = \frac{I(45°, 0°) - I(135°, 0°)}{I_3}. \tag{4.16}$$

## 4.7   The Corrections for the Measurement of the Polarization of the Fluorescence Radiation

Figure 4.15 shows the optical solid angle of the light collection system and the cone of the incident electron beam in the interaction region. From the collision region, the photons that are travelling in a direction making a non-zero angle with respect to the z-axis are also collected and focused at the cathode of the photomultiplier and this causes some depolarization of the detected light. Ehlers and Gallagher (1973) have derived a formula, for the solid angle correction, which has been used to find the necessary correction in our experiment. As shown in Fig. 4.15, if the maximum acceptance angle is $\psi_m$ for the photons, as defined by the lens used to collect the light, then the source polarization $P_s$ is linked to the observed polarization $P'_m$ by the equation

$$P'_m = \frac{(1 - \varepsilon)P_s}{(1 - \varepsilon P_s)}, \tag{4.17}$$

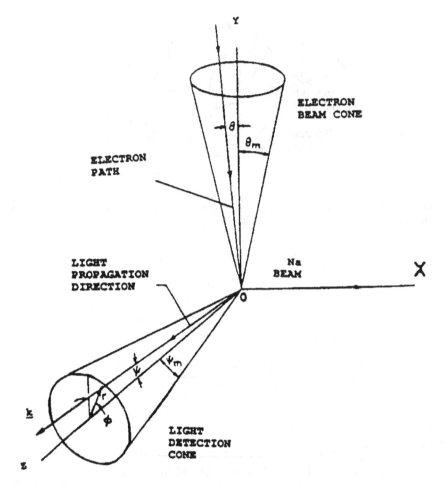

**Fig. 4.15** The coordinate system used in calculating the polarization parameters

where $\varepsilon = (1/4)\psi_m^2$. In the present experiment, $\psi_m = 0.33$ rad, therefore, $\varepsilon = 0.027$ and this correction has been applied to the present measurements.

Figure 4.15 also shows that all the electrons of the electron beam do not collide with the sodium and potassium atoms at right angle, therefore, the measured polarization of the photons is less than the source polarization $P_s$. If $\theta_m$ is the maximum acceptance angle of the electrons and $\delta = (1/4)\theta_m^2$, then the source polarization $P_s$ is given by (Ehlers and Gallagher 1973)

$$P_s = \frac{(1 - 3\delta)P}{(1 - \delta P)}, \tag{4.18}$$

where $P$ is the true polarization of the electron-excited atoms. This correction, although small, has been applied to the measured values.

A polarization analyzer can also reduce the measured value of polarization of the transmitted light if incomplete extinction of the light occurs when the polarizer axis is at a position perpendicular to the E-vector of the detected light. The measured polarization $P_m$ and the polarization observed after transmittance through the optics $P'_m$ are related by (Ehlers and Gallagher 1973)

$$P_m = \frac{k_\| - k_\perp}{k_\| + k_\perp} P'_m, \tag{4.19}$$

where $k_\|$ and $k_\perp$ are the transmittances for radiation with the electric vector parallel and perpendicular to the polarizer axis, respectively. In the present measurements for sodium D-lines, the polarizer used is HN-38, which is a polaroid film and has a transmission of 38%; and for potassium line (404.4 nm), the polarizer used is NH-PB film, which has a transmission of 30% at this wavelength. Further, these polarization analyzers give a high-extinction ratio for the appropriate wavelengths so that to a good approximation

$$P_m = P'_m. \tag{4.20}$$

## 4.8 Hanle Effect and the Depolarization of the Fluorescent Radiation

Figure 4.15 shows that the electron beam is incident in a direction along the negative $y$-axis, the atomic beam is moving along the $x$-axis and the detected radiation is travelling along the $z$-axis, therefore, a radiation, as a result of the collision, can be considered to be coming, from a dipole situated at the origin, with components along the $x$, $y$ and $z$ axes. Due to the cylindrical symmetry about the impact axis, for the unpolarized colliding partners, the measured intensities of the radiation at angles 45 and 135° should be equal and this would lead to, for the Stokes polarization parameters, $P_2 = P_3 = 0$, while the intensities along the directions parallel and perpendicular to the impact axis are not equal, i.e. the Stokes polarization parameter $P_1$ is not equal to zero.

Hanle (1924) found that the measured polarization may be affected by a magnetic field in the interaction region. Hanle (in a translation of the original article by Moruzzi 1991) made a thorough study of the polarization of the mercury resonance line at 253.7 nm in the presence of a weak magnetic field. The polarization of the emitted radiation was measured when the magnetic field was applied in different directions. Hanle found that, when the direction of the magnetic field was the same as the direction of the observation and the magnetic field was strong enough, the emitted radiation was totally unpolarized and, further, if the magnetic field was slowly reduced to zero the radiation showed a maximum polarization. The depolarization of the emitted radiation by the application of a magnetic field is known as Hanle effect. However, the polarization of the fluorescence radiation is not 100% without the applied magnetic field if the excited state is affected by fine and hyperfine structure interactions. Since excited atoms of both sodium and potassium have fine and hyperfine structures, therefore, theoretically the polarization of the emitted radiation is 12.9% at the threshold (Percival and Seaton 1958; Kleinpoppen 1969).

For sodium resonance lines 3P–3S, produced by electron impact excitation, the measured degree of polarization is 10% in a very weak magnetic field (Enemark and Gallagher 1972). Figure 4.16 shows the results of the measurement of the Stokes polarization parameter $P_1$ for the resonance lines 5P–4S of potassium as a function of the magnetic field strength near the threshold energy of 3.06 eV for the excitation of the states $5\,^2P_{3/2,1/2}$.

The Hanle effect can be explained (Corney 1977) by the classical theory of radiation. The linearly polarized fluorescence radiation emitted parallel to the impact axis can be considered to be coming from a single-damped oscillator which oscillates at an angular frequency $\omega_0$ and the electric field $E(t)$ varies as

$$E(t) = E(0)\exp\left\{-i\left(\omega_0 - \frac{i\Gamma}{2}\right)(t - t_0)\right\}j, \qquad (4.21)$$

where $\Gamma = 1/\tau$ is the damping constant and $\tau$ is the radiative lifetime of the excited atoms. The oscillating electron, in a finite magnetic field, experiences the Lorentz force

$$\mathbf{F} = -e\mathbf{v} \times \mathbf{B}, \qquad (4.22)$$

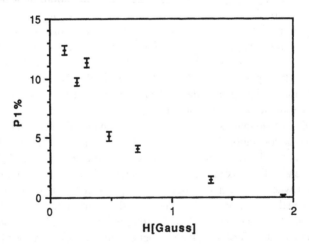

**Fig. 4.16** Measured variation in the polarization parameter $P_1$ of the potassium resonance line (5P–4S) as a consequence of the application of a magnetic field along the $z$-axis

**Fig. 4.17** Due to a strong
magnetic field or a long
lifetime of the excited state
(see text)

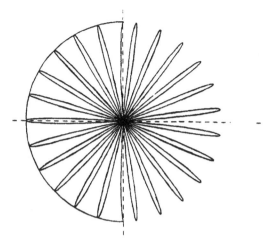

where **v** is the velocity of the electron, $e$ is the electronic charge and **B** is the magnetic inductance. The Lorentz force causes the oscillator to precess about the magnetic field with the Larmor frequency given by

$$\omega_L = g_r \frac{e}{2m} B = \frac{(g_r \mu_B B)}{\pi}, \tag{4.23}$$

where $g_r$ is the Lande g-factor to be applied for the excited $3P_{3/2}$ state of sodium and the $4P_{3/2}$ state of potassium. From (4.23), it is evident that an increase in the applied magnetic field can make the oscillator complete the rosette before the energy is radiated (see Fig. 4.17) and consequently the radiated radiation suffers more depolarization. In the case of weak applied magnetic field, the oscillator would have completed only a small fraction of the rosette (see Fig. 4.18) and the polarization of the emitted radiation would be reduced only slightly. The polarization of the emitted radiation is not only determined by the applied magnetic field but also by the lifetime of the excited state of the atom. An atom having an excited state of shorter lifetime placed in a weak magnetic field is, therefore, likely to emit more highly polarized radiation than an atom of longer lifetime excited state placed in the same magnetic field and this has lead to the use of the Hanle effect to measure the lifetimes of the excited states of atoms.

## 4.9 Fine and Hyperfine Structure of Sodium and Potassium Atoms

Sodium and potassium atoms, like all alkali metals, have a single electron outside a core of completely filled electronic shells and the outer electron has an angular momentum $L$ and an intrinsic angular momentum $S$ coupled together to produce

**Fig. 4.18** Due to a weak
magnetic field or a short
lifetime of the excited state
(see text)

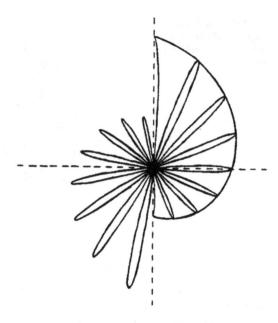

a multiple structure (called fine structure) for each energy level. Weissbluth (1978)
has given the energy shift due to the spin-orbit interaction as

$$E_{so} = \frac{Z^4 e^2 \pi^2}{4 a_o^3 m^2 o^2} \frac{J(J+1) - L(L+1) - S(S+1)}{n^3(L+1)(L+1/2)L},$$ (4.24)

where $a_o = \pi^2/me^2$ is the Bohr radius, $m$ the rest mass of the electron and $J = L + S$
is the total angular momentum and since $J$ has values of $L \pm ½$ and $S = ½$, then

$$E_{so} = E_{L\pm1/2} \frac{Z^4 e^2 \pi^2}{2 a_o^3 m^2 o^2} \frac{1}{n^3(L+1)(L+1/2)L} \left| \begin{matrix} 1/2L \\ -1/2(L+1) \end{matrix} \right.$$ (4.25)

and the splitting between $P_{1/2}$ and $P_{3/2}$ is given by

$$\delta E_{so} = 5.84 \frac{Z^2}{n^3 L(L+1)} cm^{-1}.$$

As a result the P state splits into two components as shown below where the splitting factor (Haken and Wolf 1984)

$$a = 5.84 \frac{Z^4}{n^3 L(L+1)(L+1/2)L} \, \text{cm}^{-1}. \tag{4.26}$$

An atom has, besides the angular momentum $L$ and the spin momentum $S$, the nuclear spin momentum $I$ which couples with $J$ to give the total angular momentum $F$ such that

$$F = I + J,$$

where

$$J = L + S$$

and $F$ takes the values

$$F = I + J, I + J - 1, \ldots |I - J|.$$

Since $I = 3/2$ for sodium and potassium atoms, there is further splitting of the state into hyperfine structure, e.g. the ground states $3^2S_{1/2}$ and $4^2S_{1/2}$ split into $F = 1$ and $F = 2$ states.

Kopfermann (1958) has given a general formula for calculating the hyperfine energy levels as

$$E = E_J + \frac{AC}{2} + B \frac{3/4C(C+1) - I(I+1)J(J+1)}{2I(2I-1)J(2J-1)}, \tag{4.27}$$

where $E_J$ is the energy of the multiplet level,

$$C = F(F+1) - I(I+1) - J(J-1)$$

and

$$F = I + J, I + J - 1, \ldots |I - J|. \tag{4.28}$$

Table 4.1 gives the values of hyperfine structure constants $A$ and $B$ (Happer 1975) and Fig. 4.19 shows the hyperfine energy levels of $^{23}$Na for the $3^2S_{1/2}$, $3^2P_{1/2}$ and $3^2P_{3/2}$ states.

**Table 4.1** The values of hyperfine structure constants A and B (Happer 1975)

| Element | $n$ | A($n\,^2S_{1/2}$) (MHz) | A($n\,^2P_{1/2}$) (MHz) | A($n\,^2P_{3/2}$) (MHz) | B($n\,^2P_{3/2}$) (MHz) |
|---------|-----|--------------------------|--------------------------|--------------------------|--------------------------|
| $^{23}$Na | 3 | 885.82 | 94.30 | 18.65 | 2.82 |
| $^{39}$K | 4 | 230.85 | 28.85 | 6.09 | 2.77 |
|  | 5 |  | 8.99 | 1.972 | 0.866 |
| $^{41}$K | 5 |  |  | 1.08 |  |

**Fig. 4.19** Hyperfine energy levels of $^{23}$Na for 3 $^2S_{1/2}$, 3 $^2P_{1/2}$ and 3 $^2P_{3/2}$ states, indicating the hyperfine sublevels $|FM|$. Not to scale

For $^{39}$K the hyperfine energy levels of the $4^2S_{1/2}$, $5^2P_{1/2}$ and $5^2P_{3/2}$ states are shown in Fig. 4.20 and the zero field hyperfine splittings for potassium, derived from the measured values in Table 4.1, are given below:

for $J = \frac{1}{2}$ (ground state $4^2S_{1/2}$),

$$E_{F=2} - E_{F=1} = 2A_{1/2} = 461.70 \text{ MHz},$$

for $J = \frac{1}{2}$ excited state ($5^2P_{1/2}$),

$$E_{F=2} - E_{F=1} = 2A_{1/2} = 17.98 \text{ MHz},$$

and for $J = 3/2$ excited state ($5^2P_{3/2}$)

$$E_{F=3} - E_{F=2} = 3A_{3/2} + B_{3/2} = 6.782 \text{ MHz},$$

$$E_{F=2} - E_{F=1} = 2A_{3/2} - B_{3/2} = 3.078 \text{ MHz},$$

$$E_{F=1} - E_{F=0} = A_{3/2} - B_{3/2} = 1.106 \text{ MHz}.$$

The linear polarization of the emitted radiation at threshold, according to Percival and Seaton (1958) and Kleinpoppen (1969), depends on two factors namely the fine and hyperfine structure and the natural width of the excited state. If an atom has no fine and hyperfine structure, then the level width can be ignored

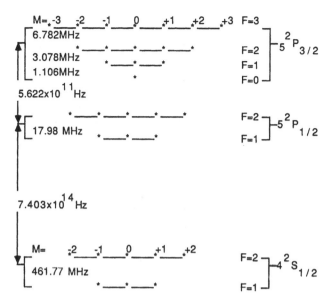

**Fig. 4.20** Hyperfine energy levels for the 4 $^2S_{1/2}$, 5 $^2P_{1/2}$ and 5 $^2P_{3/2}$ states of potassium. Not to scale

and at threshold the polarization of the P–S fluorescence following a S–P excitation is 100%, because only the $m_1 = 0$ sub-state is excited. In case the fine structure splitting is large compared with the natural level width then the polarization is 42.9%, and in the presence of hyperfine structure there is even more depolarization of the emitted radiation depending upon the value of the nuclear spin and the relation between the excited state lifetime and the hyperfine structure. When, however, the hyperfine structure splitting is large compared with the natural level width of the state, the polarization is 12.9% and in the presence of magnetic field there is further depolarization due to the Hanle effect.

## 4.10   Hanle Effect Depolarization of the Observed States in Sodium and Potassium Atoms

At the interaction region, shown in Fig. 4.21, free- and spin-polarized atoms in the atomic beam are excited by the electron collision and, as a result, resonance fluorescence light is emitted which is collected and passed through the polarization analyzer and an interference filter to the cooled photomultiplier. The 3P$_{3/2}$ state of sodium or the 5P$_{3/2}$ state of potassium, in the absence of any magnetic field, split up into four hyperfine states of total angular momentum $F$, i.e.,

$$F = I + 3/2, \; I + 1/2, \; I - 1/2, \; I - 3/2,$$

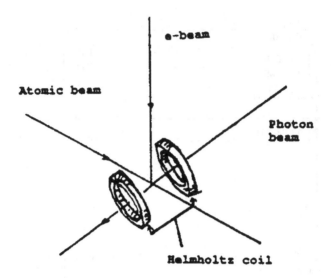

**Fig. 4.21** Helmholtz coils and the geometry of the interaction region

where $I$ is the nuclear spin. Figure 4.22 shows the relative spacing of the energy levels for $^{23}$Na and $^{39}$K (Schmieder et al. 1970). When the applied magnetic field is above zero, the $F$ states split up into their $(2F + 1)$ equidistant Zeeman levels which in total would be $(2J + 1)(2I + 1)$ Zeeman levels for the complete hyperfine structure multiplet. Since the magnetic quantum number $m_F$ can take all the integral values between $-F$ and $+F$, a $3P_{3/2}$ level at zero magnetic field would have four hyperfine structure levels ($F = 3, F = 2, F = 1$ and $F = 0$) as seen in Fig. 4.22. When a weak magnetic field is applied, the $F = 3$ level splits up into seven equidistant levels, $F = 2$ level splits up into five levels and the $F = 1$ level into three levels, while $F = 0$ level does not split. Kopfermann (1958) has given the shift in the atomic hyperfine states in a weak magnetic field $H_0$ as follows:

$$\Delta E_{F,H} = \mu_B m_F g_F H_0, \tag{4.29}$$

where

$$g_r = g_J \frac{F(F + 1) + J(J + 1) - I(I + 1)}{2F(F + 1)},$$

$$g_J = \frac{3J(J + 1) + S(S + 1) - L(L + 1)}{2J(J + 1)}$$

and $\mu_B$ = Bohr magnetron. The difference in energy between two adjacent levels is

$$\Delta\left(\Delta E_{F,H}\right) = \mu_B \Delta m_F g_F H_0$$

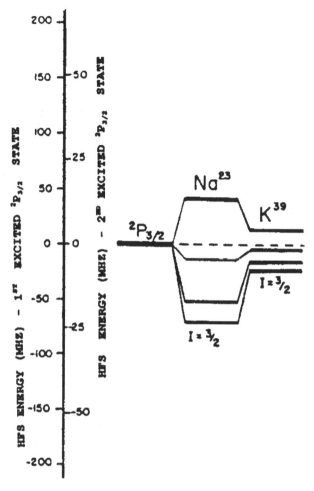

**Fig. 4.22** Relative spacing of the hyperfine energy levels of the stable alkali atoms, $^{23}$Na and $^{39}$K for the $^2P_{3/2}$ states

where $\Delta m_F = 1$, therefore,

$$\Delta\left(\Delta E_{F,H}\right) = \mu_B g_F H_0.$$

For sodium and potassium, $^2P_{3/2}$ states $I = J$, therefore,

$$g_J = \frac{4}{3} \quad \text{and} \quad g_F = \frac{2}{3}$$

and

$$\Delta\left(\Delta E_{F,H}\right) = h\nu = \left(\frac{2}{3}\right)\mu_B H_0 \tag{4.30}$$

or

$$v = \frac{(2/3)\mu_B H_0}{h},$$

where $v$ is the frequency corresponding to the energy difference of $\Delta(\Delta E_{F,H})$.

If the applied magnetic field $H_0 = 1$ G, then the atomic oscillator precesses about the field $H_0$ with a frequency of approximately 1 MHz (Sect. 4.8) and one precession takes about $10^{-6}$ s. The potassium 5 $^2P_{3/2}$ state has a lifetime of 140.8 ns (Happer 1963), therefore, the oscillator would precess through 1/7 of a complete precession equivalent to 51°. This would reduce the polarization of Stokes polarization parameter $P_1$ to below its maximum possible value and as a consequence $P_2$ would not be zero. To avoid this situation, the potassium experiment is carried out with an applied magnetic field of 0.2 G so that the oscillator precesses through 1/36 of one precession, equivalent to 10°, in the lifetime of the state and this causes only a small depolarization.

The experiment for sodium is carried out at an applied magnetic field of 0.5 G which results in 1/60 of precession in one lifetime, equivalent to 6°. Since the lifetime of 5 $^2P_{3/2}$ state of potassium is much larger than the 3 $^2P_{3/2}$ state of sodium, the linear polarization is more sensitive to the applied magnetic field for potassium than for sodium atom.

The results and discussion are described in Sect. 5.3 and the conclusion is given in Chap. 6.

## 4.11  Experimental Techniques for the Investigation of Electron Impact Excitation of Calcium and Strontium Atoms

### 4.11.1  Experimental Technique for the Study of Excitation of Calcium Atoms

In a crossed beam arrangement, the electron beam is allowed to intersect the calcium beam at right angle and the observation of excited spectral lines is made along the third orthogonal axis. The electron gun (see Sect. 3.4.5) can produce a typical electron current of 0.2 µA at lower energies of $\leq 10$ eV and $>1.0$ µA at higher energies. The calcium atomic beam is produced from an effusive stainless steel oven described in Sect. 3.5. The oven is heated to a temperature of 720°C to obtain calcium atomic beam density of $\sim 10^{11}$ atoms/cm$^3$.

Before the polarization measurements are carried out, the light emitted from the interaction region is investigated using a 0.3 m scanning monochromator (McPherson model 218) with a 2400 lines/mm grating blazed at 300 nm. The light spectrum between 250 and 450 nm is recorded at different electron energies.

Only three spectral lines in total are seen as illustrated in Fig. 4.23, a strong Ca I line at $\lambda = 422.7$ nm and two week Ca II lines at $\lambda = 393.3$ and 396.8 nm. The Ca II lines are related to the $4\,^2P_{3/2}$ and $4^2P_{1/2}$ states and show intensity ratios of approximately 2:1 at all incident electron energies.

The polarization $P$ of the emitted light is obtained by measuring the total number of photon counts with the polarizer axis parallel ($I_{\parallel}$) and perpendicular ($I_\perp$) to the electron beam axis

$$P = \frac{(I_{\parallel} - I_\perp)}{(I_{\parallel} + I_\perp)}. \tag{4.31}$$

The accuracy of the polarization measurement is checked by carrying out the measurements of the atomic resonance line $\lambda = 422.7$ nm. The data are found to agree well with those of Ehlers and Gallagher (1973).

Section 5.4.1 and Chap. 6 describe the results and discussion and the concluding remarks, respectively.

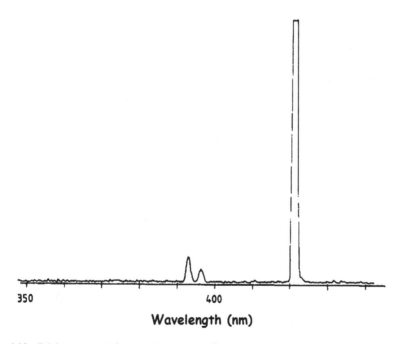

**Wavelength (nm)**

**Fig. 4.23** Calcium spectral lines excited with 40 eV electrons at an oven temperature of 700°C. The strong line recorded at $\lambda = 4222.7$ nm corresponds to the resonance transition ($4\,^1P_1 \rightarrow 4\,^1S_0$) of Ca I. The two weak lines at $\lambda = 393.3$ and 396.8 nm are related to the transitions ($4\,^2P_{3/2,1/2} \rightarrow 4\,^2S_{1/2}$) of Ca II. The intensity of the atomic resonance line has been reduced to 50% of its peak value

## 4.11.2   Experimental Technique for the Excitation of Strontium Atoms

A schematic diagram of the experimental setup is shown in Fig. 4.24. The apparatus consists of an electron gun, an atomic beam source (oven) and an electron energy analyzer, all mounted inside a vacuum chamber, and an optical detection system mounted along the tank axis outside the vacuum system. The tank is evacuated by a turbomolecular pump with a pumping speed of 500 $ls^{-1}$ and, under typical operating conditions the background pressure is about $3 \times 10^{-7}$ Torr. To reduce the influence of the Earth's magnetic field, the whole apparatus is surrounded by three orthogonal pairs of rectangular Helmholtz coils. The residual magnetic field in the region of the gun, scattering centre and electron analyzer is measured to be <30 mG, which is sufficiently low for the electron trajectories not to be affected noticeably.

The electron gun can produce an electron beam of 1 μA current and having a diameter of 8 mm at the interaction region when the incident electron energy is 45 eV. The energy of the inelastically scattered electrons is analyzed by the 127° analyzer, which is fixed to one of the platforms of a triple turntable so that the electron scattering angle can be measured with respect to the incident electron beam to within ±3°. The electron scattering angles from 30 to 150° can be used. The scattered electrons of the selected energy are detected by a channel electron multiplier.

The atomic beam is produced by effusing strontium atoms from the oven which is heated to approximately 550°C providing an estimated beam density of $10^{10}$ atoms/cm$^3$ and a beam width of 10 mm at the interaction region about 40 mm above

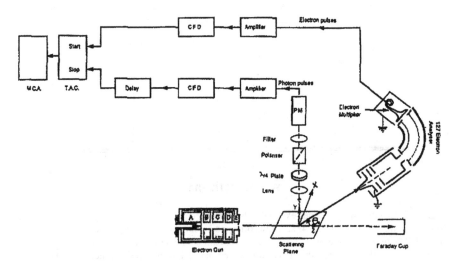

**Fig. 4.24** A schematic diagram of the experimental setup for the investigation of electron impact excitation of strontium atoms

the top of the oven nozzle. In order to obtain measurements for nearly the full range of scattering angles, the atomic beam source is tilted by 20° out of the scattering plane towards the photon detector to avoid atomic beam hitting the electron analyzer.

Light emitted from the interaction region in the $y$-direction (i.e. perpendicular to the scattering plane) is collected by a quartz lens ($f = 7.5$ cm), which is fixed 11 cm away from the centre of the vacuum chamber and forms an enlarged image of the interaction region at the cathode of the photomultiplier (EMI 9883 QB) mounted outside the vacuum system. The acceptance angle of the optical system is 0.14 srad. An interference filter, centred at 460.0 nm with a peak transmission 0f 45% and a bandwidth (FWHM) of 10.0 nm, is used to select the 460.7-nm line (5 $^1$P–5 $^1$S) and is mounted in front of the photomultiplier. The linear polarizer (Polaroid, HNP'B) is also fixed to the photomultiplier housing (in front of the interference filter) and the polarizer angle, $\alpha$, is set by rotating the photomultiplier assembly around its axis (i.e. $y$-axis) and measured from the $z$-axis towards the $x$-axis in the same way as the electron scattering angle $\theta_e$. For circular polarization measurements, a quarter-wave plate is mounted independently in front of the linear polarizer with the slow axis parallel to the electron beam.

To obtain the electron energy loss spectrum, the pass energy of the electron analyzer may be scanned with respect to the incident electron energy. The energy resolution of the present system is limited to approximately 0.9 eV (FWHM of the energy loss peaks) by the energy spread resulting from the hot filament of the electron gun. The resolution is barely sufficient to distinguish the 5 $^1$P state from higher singular states which can effect the coincidence signal of the 5 $^1$P state by cascade processes. However, the energy loss spectra show no indication of signals from higher states, the nearest being 6 $^1$S, 6 $^1$P and 5 $^1$D states that are 1.10, 1.55 and 1.64 eV above the 5 $^1$P state, respectively. The lifetime of the nearest state (6 $^1$S) is calculated to be much longer than that measured for the 5 $^1$P state, i.e. 54 ns compared with 5 ns for the 6 $^1$S state (Werij et al. 1992). The 6 $^1$P state decays predominantly to 4 $^1$D (Werij et al. 1992) and thus would contribute little. Also, as confirmed by the energy loss spectra, excitation cross section tends to drop fairly rapidly towards more highly excited states.

The overwhelming strength of the elastic peak of the energy loss spectrum still causes noticeable background at the position of the 5 $^1$P peak so that the signal-to-background ratio for 5 $^1$P electrons is only about 1.3:1. This explains the difficulties experienced in tuning the analyzer for maximum 5 $^1$P signal, since it is not easy to avoid 'tuning up' the elastic background.

For the coincidence measurements, the analyzer is set to pass electrons, which have excited the 5 $^1$P state of Sr and then have scattered through a selected angle. The scattered electrons are detected in coincidence with 5 $^1$P → 5 $^1$S decay photons emitted in a direction perpendicular to the scattering plane. The signals from both scattered electron and detected photons are fed through fast amplifiers and constant fraction discriminators, the electron signal pulse is then fed directly to the start input of the TAC and the photon signal suitably delayed (typically by 215 ns) is fed to the stop input of the TAC. An MCA is used to record the resulting pulse height (time) spectrum. The coincidence peak has an FWHM of 7 ns that includes the effect of the lifetime of the excited state, which causes a slight but noticeable tail towards larger time values. The underlying peak width given by the experimental arrangement and the electronics alone is 5–6 ns. Outputs from the discriminators are also routed to ratemeters and scalers so that the electron and photon single rates can be monitored and recorded.

The complete polarization correlation analysis requires the determination of the Stokes parameters $P_1$, $P_2$ and $P_3$ of the light emitted at right angle to the scattering plane for each electron scattering angle. The Stokes parameters have been extensively discussed in the literature (see, for example, Born and Wolf 1970) and are given by

$$P_1 = \frac{(I_0 - I_{90})}{(I_0 + I_{90})}$$

$$P_2 = \frac{(I_{45} - I_{135})}{(I_{45} + I_{135})}$$

$$P_3 = \frac{(I_{RHC} - I_{LHC})}{(I_{RHC} + I_{LHC})}$$

where $I_0$, $I_{45}$, $I_{90}$ and $I_{135}$ represent the electron-photon coincidence rates for the linear polarizer angles $\alpha = 0$, 45, 90 and 135°, respectively, measured from the incident electron direction; $I_{RHC}$ and $I_{LHC}$ are the coincidence rates for right- and left-hand circularly polarized light in the spectroscopic definition. The number of coincidences in each case is normalized to the total number of scattered electrons recorded over the duration of the run. This procedure largely removes the errors resulting from both the variation of electron beam current and the atomic beam density.

For atoms where the collision time is much shorter than any spin precession time, the electron impact excitation is expected to be fully coherent so that the excited $^1P_1$ states can be completely described by two parameters which can be derived from the measured Stokes parameters $P_1$, $P_2$ and $P_3$. Two equivalent parameter sets currently used are:

(i) The excitation parameters $\lambda$ and $\chi$ are based on the excitation amplitudes $\alpha_0$, $\alpha_{\pm1}$ in the collision coordinate system (Eminyan et al. 1974; Blum and Kleinpoppen 1975, 1979). Here $\lambda$ ($=\sigma_0/\sigma$) is the ratio of the partial cross section $\sigma_0 = |\alpha_0|^2$ for the magnetic sub-state $m_1 = 0$ to the total cross section $\sigma = |\alpha_0|^2 + 2|\alpha_1|^2$ for the excitation amplitude $\alpha_1$ and $\alpha_0$ for the magnetic sub-states with $m_1 = 1$ and $m_1 = 0$. $\alpha_0$ is chosen to be real and for symmetry reasons $\alpha_1 = -\alpha_{-1}$. For positive scattering angles, the parameters $\lambda$ and $\chi$ are linked to the measured Stokes parameters as follows (Standage and Kleinpoppen 1976; Andersen et al. 1988):

$$P_1 = 2\lambda - 1, \quad P_2 = 2[\lambda(1 - \lambda)]^{1/2}\cos\chi \quad \text{and} \quad P_3 = 2[\lambda(1 - \lambda)]^{1/2}\sin\chi.$$

(ii) The parameters that are based on the 'natural' coordinate system (Andersen et al. 1988) describe the dynamics of the excited electron charge cloud by the angular momentum transferred to the atom in the collision, $L_\perp = P_3$ and the shape of the charge cloud by the alignment angle $\gamma = \frac{1}{2}\tan^{-1}(P_2/P_1)$ and the degree of linear polarization $P_l = (P_1^2 + P_2^2)^{1/2}$ and total polarization $|P| = (P_1^2 + P_2^2 + P_3^2)^{1/2} = 1$.

Section 5.4.4 and Chap. 6 describe the results and discussion and the concluding remarks, respectively.

## 4.12   Experimental Technique for the Study of 3 $^1$P State of Helium

Figure 4.25 shows the scattering and detector coordinates commonly used to describe polarization correlation measurements (Standage and Kleinpoppen 1976; Blum and Kleinpoppen 1979). The photons are detected in the $+y$ direction ($\theta_\gamma = \phi_\gamma = \pi/2$) at right angles to the $(x, z)$-scattering plane so that in this case the scattering and detector coordinate systems are identical. A positive scattering angle (scattering

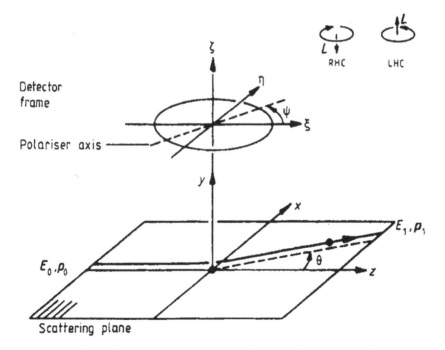

**Fig. 4.25** Scattering and detector coordinates commonly used to describe polarization correlation measurements

to the left) is shown in Fig. 4.25. This is described by $0 \leq \theta_e \leq \pi, \phi_e = 0$, while a negative scattering angle (scattering to the right) is described by $0 \leq \theta_e \leq \pi, \; \phi_e = \pi$.

For $n\,^1P$ state excitation of helium, the three scattering amplitudes $f_{ML}(\theta_e, \theta_e)$ for $M_L = 0, \pm 1$ and the phases between $f_{ML}$ can be completely described by three independent parameters (Eminyan et al. 1974):

$$\sigma = |f_0|^2 + 2[f_1]^2 \tag{4.32}$$

and

$$\lambda = \frac{[f_0]^2}{\sigma} \tag{4.33}$$

between $f_0$ (set real and positive) and $f_1$:

$$f_1 = [f_0]e^{ix}. \tag{4.34}$$

The polarization of the light measured in the coincidence experiment is conveniently described by the Stokes parameters. Usually, the relations between these quantities and $\lambda$ and $\chi$ are stated as follows (Standage and Kleinpoppen 1976; Blum and Kleinpoppen 1979):

$$P_1 = \frac{(I_0 - I_{90})}{(I_0 + I_{90})} = 2\lambda - 1, \tag{4.35}$$

$$P_2 = \frac{(I_{45} - I_{135})}{(I_{45} + I_{135})} = -2[\lambda(1 - \lambda)]^{1/2}\cos\chi, \tag{4.36}$$

$$P_3 = \frac{(I_R - I_L)}{(I_R + I_L)} = 2[\lambda(1 - \lambda)]^{1/2}\sin\chi, \tag{4.37}$$

where $I_0, I_{45}, I_{90}$ and $I_{135}$ relate to the linear polarizer angle $\psi$, as shown in Fig. 4.25 and $I_R, I_L$ correspond to right-hand and left-hand circularly polarized light. The spectroscopic definition of circularly polarized light is used (Standage and Kleinpoppen 1976). However, the helicity definition used by Blum and Kleinpoppen (1979) and Blum (1981) results in giving (4.37) an opposite sign.

Equations (4.35)–(4.37) are correct for positive scattering angles since the definition of $\chi$ (4.34) implicitly assumes $\phi_e = 0$. However, for negative scattering angles ($\theta_e, \phi_e = \pi$), (4.36) and (4.37) would return $\chi$ ($\phi_e = \pi$) $= \chi + \pi$ as a result of the underlying general relation

$$F_{M_L}(\theta_e, \phi_e) = f_{ML}(\theta_e, \phi_e = 0)e^{i(\chi + M_L\phi_e)}. \tag{4.38}$$

This apparent shift of $\chi$ by $\pi$ would contradict the definition of $\chi$ in (4.34). The inconsistency can be removed by replacing $\chi$ in (4.36) and (4.37) by

$$\chi(\phi_e) = \chi + M_L \phi_e. \tag{4.39}$$

In the case of $n$ ${}^1$P states of helium, the $M_L$ value relevant to $\chi$ is 1 so that the general relation between the Stokes parameters and $\lambda$, $\chi$ becomes

$$P_1 = 2\lambda - 1, \tag{4.40}$$

$$P_2 = -2[\lambda(1 - \lambda)]^{1/2} \cos(\chi + \varphi_e), \tag{4.41}$$

$$P_3 = 2[\lambda(1 - \lambda)]^{1/2} \sin(\chi + \varphi_e). \tag{4.42}$$

With $\phi_e = 0$ ($\pi$) for positive (negative) scattering angles.

Section 5.5 and Chap. 6 describe the results and discussion and the concluding remarks, respectively.

## 4.13   Electron Impact Excitation of 3 ${}^3$P State of Helium

In the investigation of electron impact excitation of 3 ${}^3$P state of helium, the scattered electrons and the photons are detected in coincidence (see Sect. 4.12). Electrons pass through a 127° monochromator and move along the $z$-axis to cross the helium beam emanating from a long 0.5 mm diameter capillary. Electrons that are scattered through an angle $\theta_e$ (measured towards the $+x$-axis) and have lost the energy to excite $n = 3$ states pass through a 127° analyzer and are detected by a channeltron. The photons are detected by a photomultiplier at right angles to the scattering plane in the $+y$ direction, and an interference filter for $\lambda = 389$ nm restricts the coincidence signal to the 3 ${}^3$P state. The linear polarization of the light is analyzed by a linear polarizer whose angle $\alpha$, with respect to the $z$-axis, is measured in the same way as $\theta_e$. For the circular polarization analysis, a $\lambda/4$ plate with its fast axis parallel to $z$-axis is inserted in front of the linear polarizer. The time correlation spectrum on the MCA shows a "coincidence" peak with a sharp onset determined by the time resolution of the apparatus of a few nanoseconds and a long tail determined by the lifetime of the 3 ${}^3$P state of approximately 100 ns. The signal shape may be used for a cascade-free measurement of the lifetime of the 3 ${}^3$P state (William and Humphrey 1985). A time-to-amplitude conversion range of 800 ns is used and, after subtraction of the random coincidence background, the true coincidence signal is normalized to the total number of electrons which caused valid starts in the TAC. For each value of $\theta_e$, measurements (see Sect. 4.12) are made for different values of linear polarizer angle $\alpha$.

Section 5.6 and Chap. 6 describe the results and discussion and the concluding remarks, respectively.

# Chapter 5
# Results and Discussion

**Abstract** The measurement results and discussions for the polarization of the fluorescent radiation emitted by the excited polarized atoms of sodium and potassium, the absolute cross section and polarization for the Ca II 4 $^2P_{1/2}$ state of calcium atoms, the coherence and polarization parameters for the 5 $^1P$ state in strontium atoms and the excitation of the 3 $^1P$ and 3 $^3P$ states of helium by electron impact are described. Also the measurements of partial double differential cross sections (PDDCS) and double differential cross sections (DDCS) for the ionization of helium, argon, krypton and xenon atoms and hydrogen, sulphur dioxide and sulphur hexafluoride molecules by electron impact are discussed.

**Keywords** Cross sections · Excitation · Ionization · Measurements · Polarization

## 5.1 Measurement of the Ionization Cross Sections for Atomic Gases

The electron–atom collision process,

$$e + X \rightarrow X^{n+} + ne + e_\delta,$$

where e stands for an electron and X stands for a gaseous atom, has been investigated for multiple ionization events. Detection of secondary electron (hereafter called a δ-electron, $e_\delta$) in coincidence with the product ion, $X^{n+}$, allows the identification of $n$-fold ionization process. As it is, however, not possible, in these experiments, to differentiate between the secondary or ejected electron ($e_\delta$) and the primary electron which has lost a great deal of its energy, the secondary electron must be taken to include all electrons of the selected energy.

The experimental arrangement used for these investigations is described in Sect. 4.1.1, where a typical TOF spectrum is also shown. From such a spectrum after subtraction of random coincidences the number of true coincidence events, $N_C^{(n)}$, is related to the double-differential cross-section for $n$-fold ionization, $d^2\sigma^{(n)}/(dEd\Omega)$, by

$$\frac{d^2\sigma^{(n)}}{dEd\Omega} = \frac{N_C^{(n)}}{N_i}\frac{\sigma_i}{\Delta E\Delta\Omega\varepsilon_\delta} \tag{5.1}$$

where $\sigma_i$ is the total cross-section for ion production, $N_i$ the number of detected ions and $\varepsilon_\delta$ the efficiency of the electron detection system, $\Delta E$ and $\Delta\Omega$ are the energy band width and the solid angle of the electron analyzer, respectively. For $\sigma_i$, the experimental values are taken from Schram (1966). No attempt has been made to evaluate the factor $\Delta E\Delta\Omega\varepsilon_\delta$ but it is endeavoured to keep it constant during the course of these investigations by adjusting the voltage at the ends of the channeltron in the electron analyzer so that the electrons are always incident on the channeltron with an energy of 200 eV, for maximum detection efficiency.

The double-differential cross section (DDCS) is found by summing the partial double-differential cross sections (DDCS($n+$)) over all values of $n$, that is

$$DDCS = \sum_n DDCS(n+) \tag{5.2}$$

Also the mean charge $\bar{n}(E)$, for ions detected in coincidence with the secondary electrons of energy $E$, is given by the following equation:

$$\bar{n}(E) = \frac{\left(\sum_n N_C^{(n)}n\right)}{\sum_n N_C^{(n)}} \tag{5.3}$$

### 5.1.1  Helium

Figure 5.1 shows a TOF spectrum for helium. A strong peak for $He^+$ and a very weak peak for $He^{++}$ can be seen.

Figures 5.2 and 5.3 show the values of the DDCS($n+$) and the DDCS plotted against secondary electron energy for incident electron energies of 1.0 and 2.0 keV, respectively. The present values of the DDCS have been normalized to the similar results of Opal et al. (1972) at 100 eV, secondary electron energy. Theoretical values of the DDCS using the Plane-Wave Born Approximation (PWBA) (Bell and Kingston 1975) are also shown. A common feature that can be noticed is that the curves showing the present results of the DDCS fall less rapidly than those showing results of Opal et al. (1972). Another feature that is evident in Fig. 5.3 is a small

**Fig. 5.1** TOF spectrum for helium ions He+ and He2+ detected in coincidence with ejected electrons of 100 eV energy. The angle of ejection is 90° to the incident electron direction. The incident electron energy is 2.0 keV

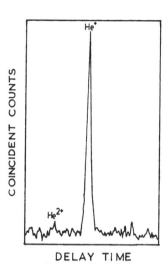

peak in the curves for the DDCS and DDCS(1+) at 35 eV, secondary electron energy, which is due, perhaps, to the auto-ionization of doubly excited He (Gibson and Reid 1986). Also in Fig. 5.3, values of DDCS(2+) show a broad peak at 100 eV, secondary electron energy. No explanation can be given for this behaviour of these results.

Figure 5.4 shows the relative values of the DDCS($n$+) for helium ions in coincidence with electrons of 200 eV secondary electron energy, plotted against incident electron energy.

### 5.1.2 Argon

Figure 5.5 shows a TOF spectrum for argon ions in coincidence with electrons of 200 eV energy when the incident electron energy is 1.5 keV and the angle of the electron ejection is 90°. Peaks for $Ar^+$, $Ar^{2+}$, $Ar^{3+}$ and $Ar^{4+}$ ions can be seen.

Figures 5.6, 5.7a and 5.8 show the relative values of the DDCS($n$+) and the DDCS plotted against secondary electron energy for incident electron energies of 1.0, 1.5 and 2.0 keV, respectively. The DDCS(1+) chiefly results from the removal of outer shell electrons, that is 3p electrons, as the probability of ionizing the 3s electrons is only 1% (Luyken et al. 1972). Multiply charged ions can result from direct multiple ionization or from inner shell ionization followed by Auger transitions (such as LMM transitions) and the shake off processes (Carlson and Nester 1973). The DDCS(2+) values show prominent peaks at 200 eV, secondary electron energy, due to the strong LMM Auger transitions in this region. The DDCS(3+) values also exhibit similar peaks but at about 20 eV lower secondary electron energy (Hippler et al. 1984a). Further, the DDCS(2+) and DDCS(3+)

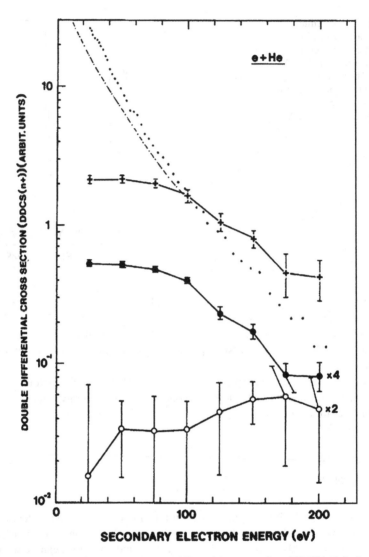

**Fig. 5.2** The measurements of partial doubly differential cross sections (DDCS(+)) for ionization of helium are plotted against the secondary electron energy, the incident electron energy being 1.0 keV. Present measurement values of DDCS(1+) denoted by (•) refer to He+ ion and DDCS(2+) values denoted by (o) refer to He2+ ion. Also shown are the present measurements of doubly-differential cross section (DDCS) for ionization denoted by (+) while (.. . ..) refers to the results of Opal et al. (1972). The present measurements of DDCS(1+) and DDCS(2+) shown have been divided by 4 and 2, respectively. (-.-.-.-) refer to the theoretical values of Bell and Kingston (1975). Lines are to guide the eye

also show broad peaks at 40 eV which could be due to the Coster–Kronig transitions (Mehlhorn 1968). Figure 5.7b shows the variation of mean charge $\bar{n}(E)$ with secondary electron energy. It is evident that low energy secondary

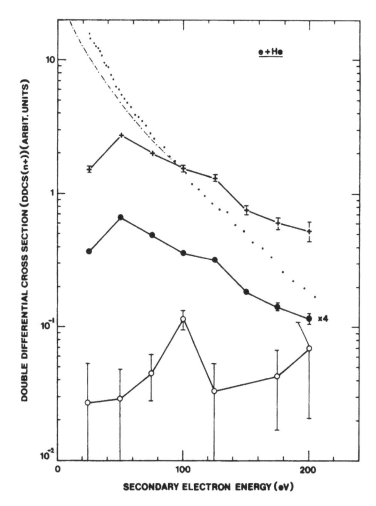

**Fig. 5.3** Relative values of DDCS($n+$) and DDCS are plotted against secondary electron energy, the incident electron energy being 2.0 keV. (+) refers to the present measurements of DDCS which have been normalized to the results of Opal et al. (1972), denoted by (.....), at 100 eV secondary electron energy. Present measurement values of DDCS(1+) denoted by (●) refer to He+ ion and the DDCS(2+) values denoted by (o) refer to He2+ ion. For better presentation the present measurements of DDCS(1+) have been divided by 4. (-.-.-.-) refers to the theoretical values of Bell and Kingston (1975). The lines are to guide the eye

electrons are mostly associated with single ionization while detected electrons with about 200 eV energy are mostly associated with the multiply charged ions and $\bar{n}(E)$ has values higher than two. In fact, the peak at 185 eV due to LMM Auger transition is quite evident here.

Figure 5.9 shows the relative values of the DDCS plotted against secondary electron energies of 1.0, 1.5 and 2.0 keV. Similar results obtained by Opal et al. (1972) for 500 eV, incident electron energy, are also shown for comparison. Present results seem to agree generally with the results of Opal et al. (1972).

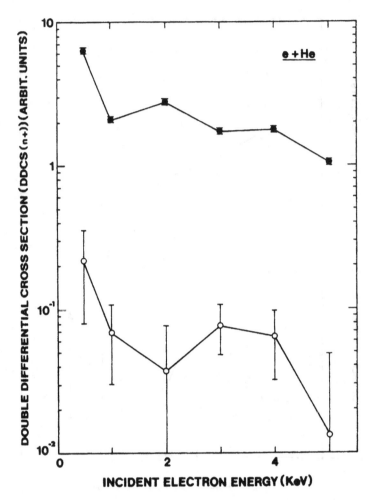

**Fig. 5.4** The present measurements of DDCS($n+$) are plotted against the incident electron energy for secondary electron energy of 200 eV. Present measurement values of DDCS($1+$) denoted by (*closed circles*) refer to He+ ion and DDCS($2+$) values denoted by (*open circles*) refer to He2+ ion

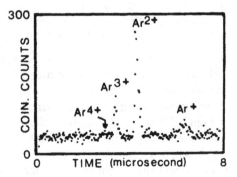

**Fig. 5.5** A TOF spectrum for argon ions in coincidence with secondary electrons

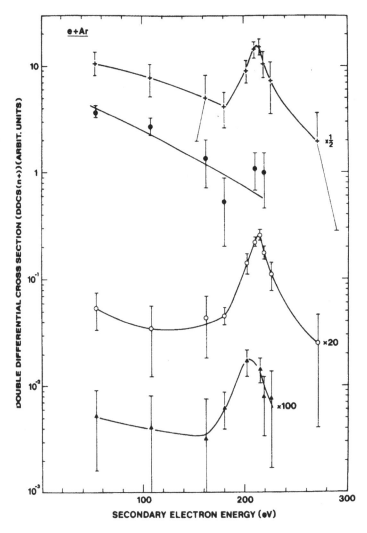

**Fig. 5.6** Present measurements of DDCS($n+$) and DDCS are plotted against secondary electron energy, incident electron energy being 1.0 keV; (*closed circles*) refer to Ar+, (*open circles*) refer to Ar2+ and (*closed triangles*) refer to Ar3+ ions. (*Plus symbols*) refer to the present measurements of DDCS. Present values of DDCS(2+), DDCS(3+) and DDCS have been divided by 20, 100 and 0.5, respectively, for better presentation. Lines are to guide the eye

Figure 5.10 shows the relative values of DDCS($n+$) plotted against incident electron energy, for secondary electrons having 200 eV energy. At this ejected electron energy multiple ionization results mainly from $L$-shell ionization, whereas singly charged ions result from $M$-shell ionization. According to Krause (1979), 93.4% of the vacancies in the $L_1$-subshell can decay by the Coster–Kronig transitions resulting in $L_2$ or $L_3$ vacancies and these vacancies then, almost exclusively,

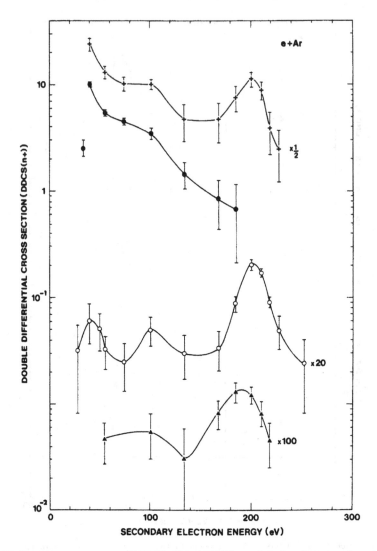

**Fig. 5.7** Present measurements of DDCS($n+$) and DDCS are plotted against secondary electron energy, incident electron energy being 1.5 keV; (*closed circles*) refer to Ar+, (*open circles*) refer to Ar2+ and (*closed triangles*) refer to Ar3+ ions. (*Plus symbols*) refer to the present measurements of DDCS. Present values of DDCS(2+), DDCS(3+) and DDCS have been divided by 20, 100 and 0.5, respectively, for better presentation. Lines are to guide the eye

decay via Auger electron emission. In addition, the shake-off processes, following $L$-vacancy production, have a probability of occurrence of about 15% (Carlson and Nester 1973). The calculated values of $\sigma(L)$ (Hippler et al. 1984b) for the production of different charge states after the creation of $L$-vacancies, are compared with the present results for an incident electron energy of 2.0 keV in the Table 5.1. The results of Stolterfoht et al. (1973) are also given in the Table 5.1 for comparison.

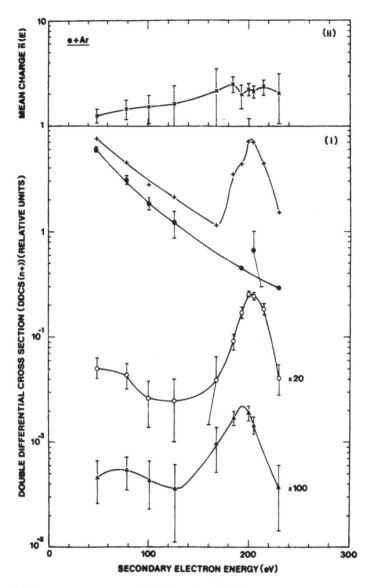

**Fig. 5.8** (a) Present measurements of DDCS($n+$) and DDCS are plotted against the secondary electron energy, the incident electron energy being 2.0 keV; (*closed circles*) refer to Ar+, (*open circles*) refer to Ar2+ and (*closed triangles*) refer to Ar3+ ions. (*Plus symbols*) refer to the present measurements of DDCS. Present values of DDCS(2+) and DDCS(3+) have been divided by 20, 100, respectively, for better presentation. (**b**) Shows the present measurement values of the mean charge per ion at the incident energy $E$ (i.e. $\bar{n}(E)$) as a function of the secondary electron energy, the incident electron energy being 2.0 keV. The lines are to guide the eye

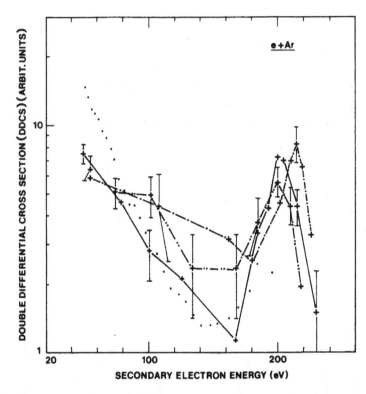

**Fig. 5.9** Present measurements of DDCS are plotted against secondary electron energy for different incident energies; (——) refers to the incident electron energy of 2.0 keV, (-..-..-) refers to the incident electron energy of 1.5 keV and (-.-.-) refers to the incident electron energy of 1.0 keV and (......) refers to the similar results of Opal et al. (1972) for an incident electron energy of 500 eV. Present measurement values of DDCS for 100 eV secondary electron energy and 2.0 keV incident electron energy have been normalized to the results of Opal et al. (1972)

Figure 5.11 shows the values of the DDCS($n+$) plotted against the incident electron energy for a secondary electron energy of 200 eV. Assuming that the Auger emission is isotropic and neglecting direct-double ionization of $M$-shell, the DDCS(2+) and DDCS(3+) are given by (Van der Wiel and Wiebes 1971; Hippler et al. 1984b) the following equations:

$$\text{DDCS}(2+) = 0.89\,\sigma\,(L_{23}), \tag{5.4}$$

$$\text{DDCS}(3+) = 0.84\,\sigma\,(L_1) + 0.92\,\sigma\,(L_{23})\,S, \tag{5.5}$$

where $S$ is the Shake-off probability following $L$-shell ionization. Equation (5.4) gives good agreement with the DDCS(2+) data except at energies below 2.0 keV (Hippler et al. 1984b) and (5.5) with $S = 0$ also agrees fairly well with the present data.

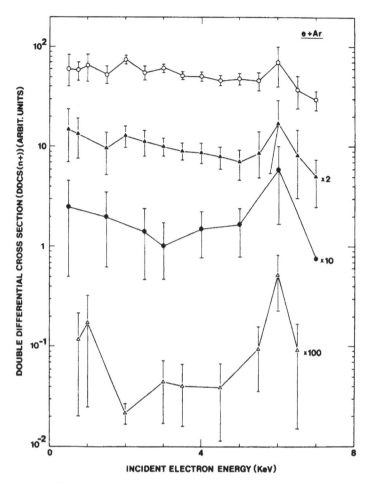

**Fig. 5.10** Present measurements of DDCS($n+$) are plotted against incident electron energy for secondary electron energy of 200 eV; (*closed circles*) refer to Ar+, (*open circles*) refer to Ar2+, (*closed triangles*) refer to Ar3+ and (*open triangles*) refer to Ar4+ ions. Present values of DDCS (1+), DDCS(3+) and DDCS(4+) have been divided by 10, 2 and 100, respectively, for better presentation. Lines are to guide the eye

**Table 5.1**

| Charge state | Present Results | $\sigma(L)$ | Stolterfoht et al. (1973) |
|---|---|---|---|
| 2+ | 72.5 | 70.7 | 54 |
| 3+ | 25.2 | 26.8 | 44 |
| 4+ | 2.3 | 2.5 | 2 |

Figure 5.12 shows the relative values of the DDCS($n+$) for $n = 1$–4 and of the total doubly differential cross section as a function of the angle of ejection of the secondary electrons of 200 eV energy for an incident electron energy of 1.0 keV.

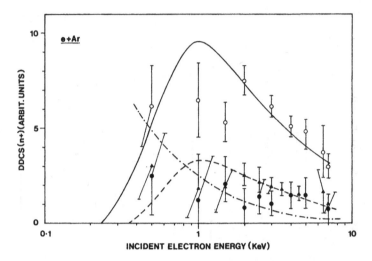

**Fig. 5.11** Present measurements of DDCS($n+$) are plotted against incident electron energy for secondary electron energy of 200 eV; ($\bullet$) refers to Ar+, (o) refers to Ar2+ and ($\blacktriangle$) refers to Ar3+ ions; (————) and (- - - -) refer to DDCS(2+) and DDCS(3+), respectively, for the $L$-shell contributions with $S = 0$; (-.-.-.-) refers to the singly-differential cross section for $M$-shell ionization

**Fig. 5.12** Present measurements of DDCS($n+$) and DDCS as a function of the angle of ejection of a secondary electron with an energy of 200 eV, incident electron energy being 1.0 keV; (*closed circles*) refer to Ar+, (*open circles*) refer to Ar2+, (*closed triangles*) refer to Ar3+ and (*open triangles*) refer to the Ar4+ ions. (*Closed squares*) refer to the present measurements of DDCS and (*multiplication symbols*) refer to the experimental results of Opal et al. (1972). The lines are to guide the eye

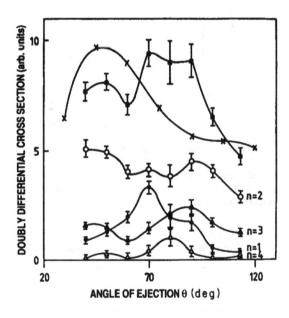

The figure also shows the measurements of Opal et al. (1972) for the total doubly differential cross section for an incident electron energy of 500 eV. The present measurements of DDCS(1+) show a pronounced maximum at an angle of 70°, which is close to the maximum expected at about 63° (Ogurtsov 1973; Lahman-Bennani et al. 1983) for a binary-type collision between the incident and the atomic electrons. The DDCS(2+) is known to include not only the effect of the ions produced by inner shell ionization followed by Auger electron emission but also the ions resulting from other processes such as double ionization (Mathis and Vroom 1976). The present measurements of the DDCS(2+) show some anisotropy, which can possibly be attributed to the ejected electrons resulting from the Auger transitions such as $L_3 M_{23} M_{23}$ (Cleff et al. 1974; Berezhko et al. 1978; Sewell and Crowe 1982). The DDCS(3+) and DDCS(4+) can be seen to exhibit a marked anisotropic behaviour, which could be due to an anisotropic angular distribution of Auger electrons emitted in the presence of one or two additional vacancies or satellite Auger electrons (Mehlhorn 1985; McGuire 1975). This could perhaps be the first evidence of anisotropy in the emission of satellite Auger electrons such as $L_3 M - MMM$, for example, which may have an anisotropic angular distribution similar to $L_3 MM$ Auger electrons.

The present measurements of the total doubly differential cross section when compared with the measurements of Opal et al. (1972) at 500 eV show a difference in shape due partly to the expected shift in the binary collision peak resulting from the higher incident energy of 1 keV used here and also to the production of more highly charged ions at higher incident electron energy.

### 5.1.3 Krypton

Figure 5.13 shows a TOF spectrum for krypton ions detected in coincidence with the secondary electrons having an energy of 85 eV, incident electron energy being 10.0 keV. Peaks for ions having charge states 1+ to 6+ can be seen. As krypton has several isotopes, each peak is a sum of several closely spaced peaks, one for every isotope (Short et al. 1986). The inset in the Fig. 5.13 is a magnified image of the

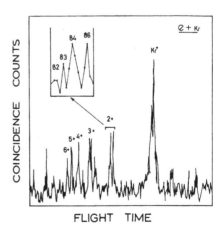

**Fig. 5.13** A TOF spectrum for krypton. The inset shows the isotopic spread of the peak for Kr2+

peak for $Kr^{2+}$ ion to show more clearly the separate peaks for different isotopes of krypton.

Figure 5.14 shows TOF spectra for different secondary electron energies, the incident electron energy being the same. This demonstrates the variation in the relative strength of the peaks for the ions of different charge states as the energy of the accompanying secondary electron changes.

Figure 5.15 shows relative values of the DDCS($n+$) and DDCS plotted against secondary electron energy. The incident electron energy is 10.0 keV and the angle of ejection is 90°. Present values of DDCS are found to agree generally with similar results of Opal et al. (1972) for incident electron energy of 500 eV. The small difference could well be due to the difference in incident electron energies. The shape of the DDCS curve is also found to agree with the results of Oda et al. (1972) (not shown in the figure). The following features can also be noted about the DDCS ($n+$) values.

1. The DDCS(2+) has relatively higher values at lower secondary electron energies except around 200 eV, where it shows an increase. Prominent peaks can also be seen in the regions of the Coster–Kronig transitions LMM, MMN and the Auger transitions MNN.
2. The DDCS(3+) has its highest value at the lowest energy measured for the detected electron. It shows a monotonic decrease with detected electron energy except for the peaks near the strong transitions such as $L_1 L_3 M_{45}$, $M_1 M_4 M_{45}$ and $M_{23} M_{45} N_{23}$.
3. The DDCS (4+) also shows higher values at lower secondary electron energies. There is a prominent peak around 70 eV which is about 15 eV lower than the corresponding peak in the DDCS(3+). This agrees with the results of Hippler et al. (1984a) who found a shift of 20 eV per charge in similar peaks.

Figure 5.16 shows relative values of the DDCS($n+$) plotted against incident electron energy, secondary electron energy being 60 eV. The DDCS($n+$) generally shows changes in their values at $L$-sub-shell ionization potentials. The DDCS(5+) shows a prominent peak at the $L_{23}$-sub-shell ionization potentials.

Figure 5.17 shows the values of DDCS($n+$)/DDCS(1+), from the data of Fig. 5.16, plotted against the incident electron energy. The variations in the values of the ratios, DDCS($n+$)/DDCS(1+), at $L$-shell edges are evident.

## 5.1.4  Xenon

Figure 5.18 shows a TOF spectrum for xenon ions detected in coincidence with the secondary electrons of 30 eV energy when the incident electron energy is 6.0 keV. Peaks for xenon ions having charge states from 1+ to 9+ can be seen. The width of the peaks is due mainly to the isotopic spread of xenon ions, the contribution due to the thermal energy spread of the gas atoms being much less in comparison (Short et al. 1986). The asymmetry in the shape of the peaks is due to the unequal isotopic abundance in xenon.

**Fig. 5.14** TOF spectra for krypton ions for secondary electron energies of (**a**) 15 eV, (**b**) 63.5 eV and (**c**) 70.0 eV. The incident electron energy is 10.0 eV

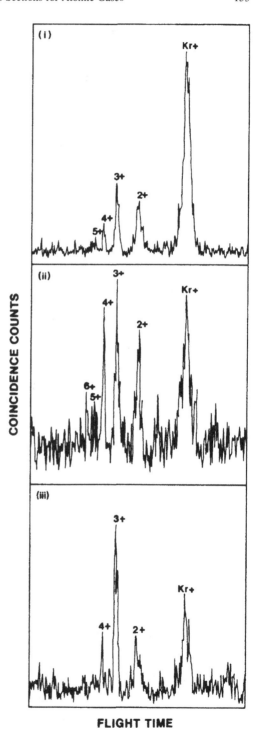

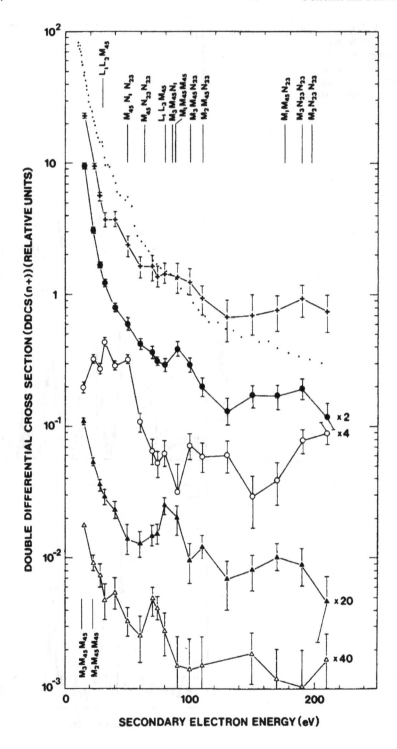

SECONDARY ELECTRON ENERGY (eV)

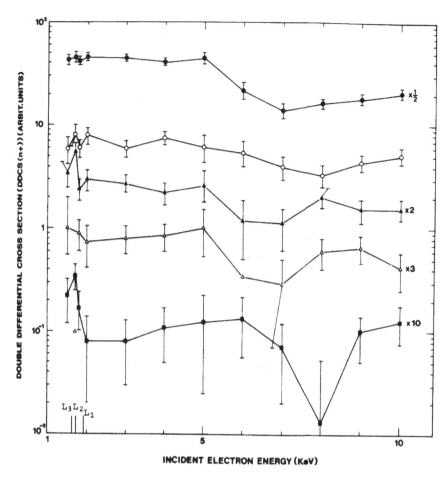

**Fig. 5.16** Present measurements of DDCS($n+$) as a function of the secondary electron energy, the incident electron energy being 10.0 keV; (*closed circles*) refer to Kr+, (*open circles*) refer to Kr2+, (*closed triangles*) refer to Kr3+, (*open triangles*) refer to the Kr4+ and (*closed squares*) refer to Kr5+ ions. The DDCS(1+), DDCS(3+), DDCS(4+) and DDCS(5+) have been divided by ½, 2, 3, and 10, respectively. The lines are to guide the eye

Figure 5.19a shows the relative values of the DDCS($n+$) and the DDCS plotted against secondary electron energy, incident electron energy being 6.0 keV. The DDCS generally exhibits a smooth decrease (Oda et al. 1972; Opal et al. 1972;

**Fig. 5.15** Present measurements of DDCS($n+$) and DDCS as a function of the secondary electron energy, the incident electron energy being 10.0 keV; (*closed circles*) refer to Kr+, (*open circles*) refer to Kr2+, (*closed triangles*) refer to Kr3+ and (*open triangles*) refer to the Kr4+ ions. (*Plus symbols*) refer to the present measurements of DDCS and (*dotted line*) refer to the measurements of Opal et al. (1972). The present values of the DDCS have been normalized to the data of Opal et al. (1972) at 85 eV, secondary electron energy. The lines are to guide the eye

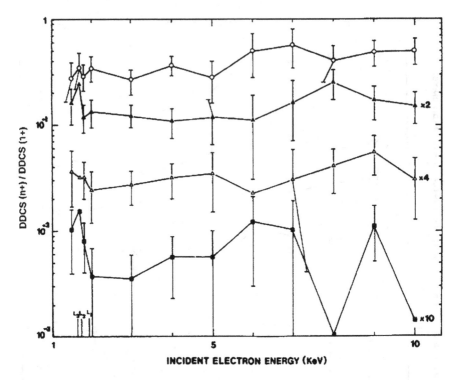

**Fig. 5.17** DDCS($n+$)/DDCS($1+$) is plotted against the incident electron energy, secondary electron energy being 60 eV. (*Closed circles*) refer to Kr+, (*open circles*) refer to Kr2+, (*closed triangles*) refer to Kr3+, (*open triangles*) refer to the Kr4+ and (*closed squares*) refer to Kr5+ ions. The lines are to guide the eye

Ogurtsov 1973) with increase in the energy of the emitted electrons. Electrons ejected from auto-ionization states have definite sharp energies and the corresponding spectrum is superimposed on the spectrum due to the direct ionization and the double Auger (Mehlhorn 1985) transitions such as $N_{45}OOO$ which accounts for 27% of the total radiation-less transition rate (Cairns et al. 1969) from $N_{45}$ subshells. The sharp increase in the DDCS corresponds to the Auger transition $N_{45}$ $O_1O_{23}$ (McGuire 1982) emitting 32.8 eV electrons. Figure 5.14a also shows that the present results for the DDCS agree generally with the similar results of Opal et al. (1972) for 500 eV, incident electron energy. At secondary electron energies greater than 100 eV, however, the values of the DDCS are higher than those of Opal et al. (1972) and this increase could be due to the several additional single (Coghlan and Clausing 1973) and double (Mehlhorn 1985) Auger transitions, which are possible at 6.0 keV incident electron energy.

Figure 5.19b shows the mean charge ($\bar{n}\,E$) as a function of the secondary electron energy $E$. The higher mean charge values around 40 eV are due perhaps to the strong $N_{45}\,O_1\,O_{23}$ Auger transitions (Coghlan and Clansing 1973) in this region. As

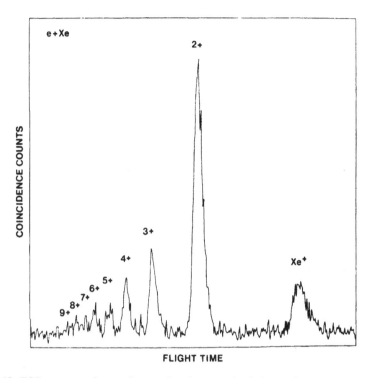

**Fig. 5.18** TOF spectrum of xenon ions produced as a result of electron impact on xenon atoms. Incident electron energy and secondary electron energies are 6.0 keV and 30 eV, respectively. The electron ejection angle is 90°. Peaks for Xe1+ to Xe9+ can be seen

a result, perhaps, of the absence of Auger electrons in the region of 100 eV secondary electron energy, the value of $\bar{n}\ (E)$ shows a minimum.

Figure 5.20 shows the relative values of the DDCS($n+$) for $n = 1$–8 plotted against the incident electron energy for an ejected electron energy of 30 eV and an ejection angle of 90°. Higher order ionization is mainly produced by the removal of inner-shell electrons followed by vacancy cascade. A vacancy in the $L_1$-sub-shell, for example, has a 52.4% possibility (Chen et al. 1981) of being transferred to the $L_{23}$-sub-shells by a Coster–Kronig transition which could be followed by several Auger/Coster–Kronig transitions (Coghlan and Clausing 1973) such as $L_3\ M_3\ M_3$, $M_3\ M_5\ O_{45}$, $N_5\ N_{45}\ N_{45}$ and $N_{45}\ O_1\ O_{23}$, resulting in the ejection of up to five or more electrons. If one adds to this the contributions due to shake-off (Carlson and Nester 1973) processes as well, the high state of ionization reached in these processes is quite understandable. Figure 5.20a also shows that, at incident electron energies higher than $M$-shell ionization potential, the DDCS($n+$) for all charges show an increase in their values. The DDCS(2+) and DDCS(4+) have peaks at 4.0 keV, while the DDCS(3+) has a peak at 3.0 keV incident electron energy. The DDCS(2+), DDCS(3+) and DDCS(4+) show jumps in their values at the $L$-sub-shell edges. A trend towards production of ions with higher charges, beyond

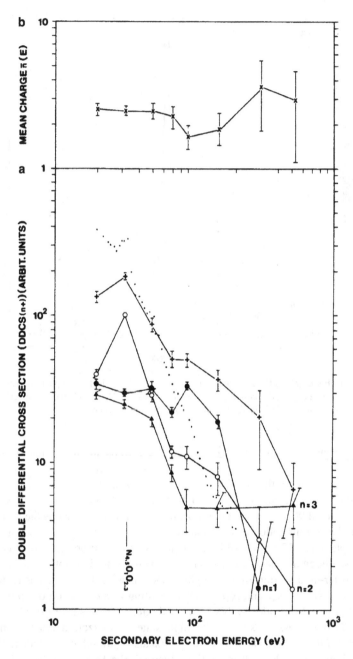

**Fig. 5.19** (a) Present measurements of DDCS($n+$) and DDCS as a function of the secondary electron energy, the incident electron energy being 6.0 keV; (*closed circles*) refer to Xe+, (*open circles*) refer to Xe2+ and (*closed triangles*) refer to Xe3+ ions. (*Plus symbols*) refer to the present values of DDCS. (*Dotted line*) refers to the results of Opal et al. (1972) for the DDCS at 500 eV, incident electron energy. (**b**) Shows the values of mean charge (i.e. $\bar{n}(E)$) as a function of the secondary electro energy, incident electron energy being 6.0 keV. The lines are to guide the eye

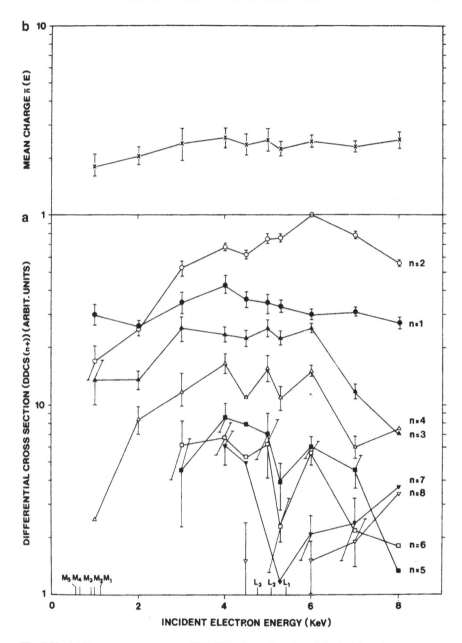

**Fig. 5.20** (a) Present measurements of DDCS($n+$) as a function of the incident electron energy, the secondary electron energy being 30 eV; (*closed circles*) refer to Xe+, (*open circles*) refer to Xe2+, (*closed triangles*) refer to Xe3+, (*open triangles*) refer to Xe4+, (*closed squares*) refer to Xe5+, (*open squares*) refer to Xe6+, (*inverted closed traingles*) refer to Xe7+, and (*inverted open triangles*) refer to Xe8+ ions. (**b**) Shows the values of mean charge (i.e. $\bar{n}\ (E)$) as a function of the incident electro energy, secondary electron energy being 30 eV. The lines are to guide the eye

7.0 keV, incident electron energy, is also evident. This observation agrees generally with the results of photoionization studies (Short et al. 1986; Carlson et al. 1966) showing that vacancies in the $L$-shell can result in a higher average charge per ion than vacancies in the $M$-shell or $N$-shell.

Figure 5.20b shows the values of $\bar{n}(E)$ for $E = 30$ eV plotted against the incident electron energy, and reveals that there is a relative increase in production of ions with higher charges at incident electron energies greater than 1.0 keV. Jumps in the value of $\bar{n}(E)$ for energies just above the $L$-shell edges are also evident.

### 5.1.5 Neon

Figure 5.21 shows a TOF spectrum for neon ions, detected in coincidence with secondary electrons of energy 200 eV, incident electron energy being 3.0 keV. Neon has two isotopes $Ne^{20}$ and $Ne^{22}$ having 90.8 and 8.8% abundances, respectively. There is a 10% difference in their masses which is much larger than the mass resolution of the ion-analyzer, therefore, the peaks for each charge state show a break up into two separate peaks. The spectrum shows peaks for $^{22}Ne^{+}$, $^{20}Ne^{+}$, $^{22}Ne^{2+}$, $^{20}Ne^{2+}$, $^{20}Ne^{3+}$ and $^{20}Ne^{6+}$. For higher charge states the peaks become smaller and for $^{22}Ne$ isotope, being less abundant, become too small to be identified. The peak for $^{20}Ne^{+}$ is most prominent. The ratio for DDCS(2+)/DDCS(1+) has been estimated to be 0.055 ($\sim$40% error).

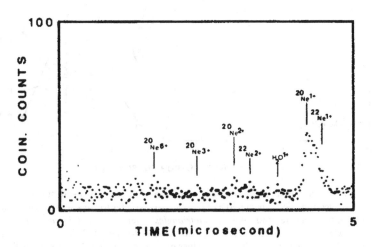

**Fig. 5.21** TOF spectrum for neon ions detected in coincidence with secondary electrons of 200 eV. Incident electron energy is 3.0 keV

## 5.1.6 X-Ray–Ion Coincidence Measurements

Figure 4.6 shows the experimental arrangement for the X-ray–ion coincidence study carried out for xenon atoms under electron impact. In this investigation xenon ions produced as a result of the ionization are detected in coincidence with the X-rays emitted by the ion at 90° to the incident electron direction. The X-rays are detected by a liquid-nitrogen-cooled HPGe detector and the ions are analyzed, in respect of their charge state, by a TOF type ion analyzer described in Sect. 3.1.6.Figure 4.7 shows the electronic circuit used to record coincidences between the detected ions and the X-rays. Figure 5.22 shows such a TOF spectrum for xenon ions detected in coincidence with the emitted X-rays as a result of the impact, on xenon atoms, of electrons having 11.0 keV energy. The TOF spectrum shown in Fig. 5.22 has been used to find the relative abundances of xenon ions, having different charge states, which are detected in coincidence with the emitted X-rays. Figure 5.23 shows that $Xe^{8+}$ has the highest intensity amongst the detected ions. Present results yield an average charge per ion, detected in coincidence with the emitted X-rays, of 6.9. This distribution of charge state can result (Carlson et al. 1966; Short et al. 1986) from the $L$-shell ionization followed by the Coster-Kronig transitions (Chen et al. 1981), a radiation transition and several possible Auger transitions (Coghlan and Clausing 1973). The contribution of the shake-off process towards increasing the state of ionization may also not be negligible (Carlson and Nester 1973).

## 5.1.7 X-Ray Spectroscopy Using a Crystal X-Ray Spectrometer

Figure 5.22 shows a second order spectrum of the $K_\alpha$ and $K_\beta$ lines of $^{54}$Mn taken by using a Nacl(100) crystal. These X-rays are emitted by a $^{55}$Fe radioactive source as a result of electron capture (EC) (Hatch 1964; Schnopper 1966). The experimental

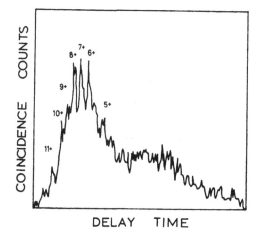

**Fig. 5.22** TOF spectrum of xenon ions detected in coincidence with X-rays emitted at 90° to the incident electron direction. The incident electron energy is 11.0 keV

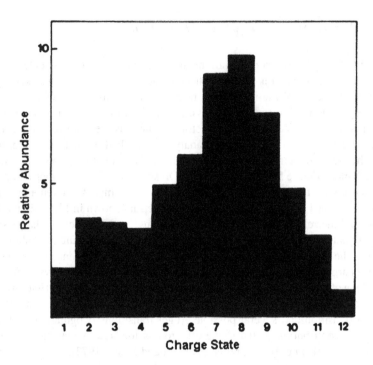

**Fig. 5.23** Relative abundance of xenon ions as a function of the charge state of ions. The xenon ions are detected in coincidence with X-rays emitted at 90° to the incident electron direction. The incident electron energy is 11.0 keV

arrangement and the electronic set-up is given in Sect. 4.3. From Fig. 5.24 the spectrometer resolution for $K_\alpha$ line is found to be 161 eV.

A comparison between Figs. 5.24 and 5.25 shows that the crystal X-ray spectrometer has over ten times better resolution than the HPGe detector (Wille and Hippler 1986). This crystal X-ray spectrometer can, therefore, be used in X-ray spectroscopy in the keV energy range with a considerable advantage, provided sufficient intensity of X-rays is available.

## 5.2 Measurement of the Ionization Cross Sections for Molecular Gases

### 5.2.1 Hydrogen

The partial doubly differential cross sections (PDDCS) for the production of $H^+$ and $H_2^+$ and the doubly differential cross sections (DDCS) for the ionization of $H_2$ have been measured for the incident electron energy ($E_i$) of 100 eV, ejected electron energies ($E_\delta$) of 20, 30, 40, 60 and 80 eV and $\theta_\delta$ from 2 to 110°.

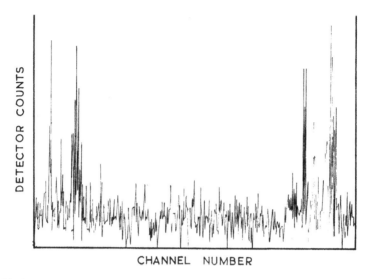

**Fig. 5.24** 54Mn k-emission spectrum from 55Fe electron capture (EC) X-ray source as recorded by the crystal X-ray spectrometer using NaCl (100) crystal

**Fig. 5.25** 54Mn $K_\alpha$ and $K_\beta$ line spectrum as recorded by the HPGe detector

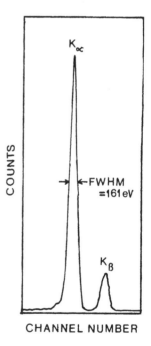

A typical TOF spectrum of $H_2$ is shown in Fig. 5.26. Figures 5.27–5.30 show the measurements of DDCS and PDDCS($H^+$) as a function of $\theta_\delta$ for $E_\delta$ having values of 20, 30, 40, and 60. Figure 5.31 shows measurements of the DDCS for $E_\delta$ equal to 80 eV. Figure 5.32 shows the DDCS and PDDCS($H^+$) as a function of $\theta_\delta$ for $E_\delta$

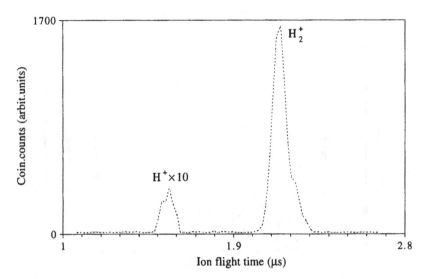

**Fig. 5.26** TOF spectrum for hydrogen ions detected in coincidence with secondary electrons having energy of 60 eV. The incident electron energy is 100 eV and the angle of electron ejection ($\theta_\delta$) is equal to 20°

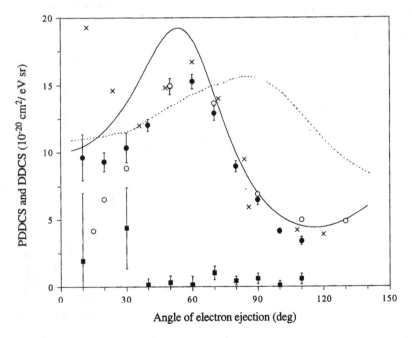

**Fig. 5.27** Present measurements of partial doubly differential cross sections (PDDCS) for ionization and doubly differential cross sections (DDCS) as a function of the angle of electron ejection ($\theta_\delta$). The secondary electron energy ($E_\delta$) is equal to 20 eV and the incident electron energy ($E_i$) is equal to 100 eV. (*Closed squares*) refer to PDDCS for H+ ion and (*closed circles*) refer to the DDCS. The figure also shows the published experimental values of Shyn et al. (1981) (denoted by *multiplication symbols*) and DuBois and Rudd (1978) (denoted by *open circles*) and the results of Rudd (1991) (denoted by *line*) and Schultz et al. (1992) (denoted by *dotted line*)

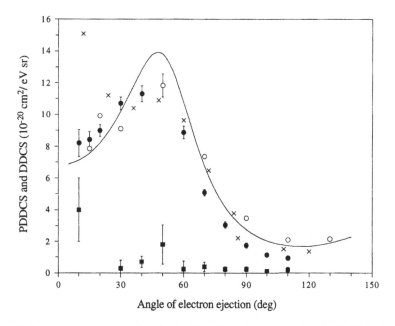

**Fig. 5.28** Present measurements of partial doubly differential cross sections (PDDCS) for ionization and doubly differential cross sections (DDCS) as a function of the angle of electron ejection
($\theta_\delta$). The secondary electron energy ($E_\delta$) is equal to 30 eV and the incident electron energy ($E_i$) is
equal to 100 eV. (*Closed squares*) refer to PDDCS for H+ ion and (*closed circles*) refer to the
DDCS. The figure also shows the published experimental values of Shyn et al. (1981) (denoted by
*multiplication symbols*) and DuBois and Rudd (1978) (denoted by *open circles*) and the results of
Rudd (1991) (denote by *line*)

having a value of 20 eV and $E_i$ equal to 150 eV. Figures 5.27 and 5.32 also show
for comparison the experimental data of DuBois and Rudd (1978) and Shyn et al.
(1981), and the theoretical data of Rudd (1991) and Shultz et al. (1992) for
the DDCS for the ionization of $H_2$. Present measurements for $E_i = 100$ eV have
been normalized to the data of DuBois and Rudd (1978) at $\theta_\delta = 50°$ and the data for
$E_i = 150$ eV have been normalized to those of Shyn et al. (1981) at $\theta_\delta = 60°$.

Figures 5.27–5.31 show that, in general, the values of PDDCS(H$^+$) are much
lower than those of PDDCS(H$_2$$^+$), therefore, the form of curves for DDCS is not
significantly affected by the presence of the dissociative single ionization. The
DDCS values in Fig. 5.27 show an agreement with the published data of DuBois
and Rudd (1978) and Shyn et al. (1981) for angles higher than 40°, with those of
Rudd (1991) at $\theta_\delta > 70°$ and with those of Schultz et al. (1992) at $\theta_\delta < 30°$. There
are, however, significant disagreements for $\theta_\delta$ higher than 60° with those of Shultz
et al. (1981) and for $\theta_\delta$ lower than 40° with those of Shyn et al. (1981) and DuBois
and Rudd (1978). Figure 5.27 also shows that while the data of Shyn et al. (1981)
has a sudden rise at lower ejection angles, the data of DuBois and Rudd (1978)
descends steeply towards the lower ejection angles, although, present DDCS values
remain nearly constant at the lower ejection angles thus showing some agreement
with the classical description of Shultz et al. (1992). Figure 5.28 shows that the

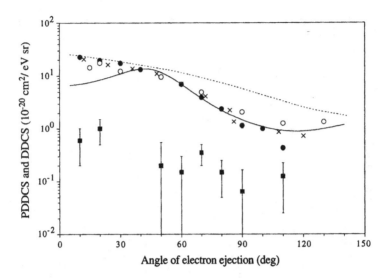

**Fig. 5.29** Present measurements of partial doubly differential cross sections (PDDCS) for ionization and doubly differential cross sections (DDCS) as a function of the angle of electron ejection
($\theta_\delta$). The secondary electron energy ($E_\delta$) is equal to 40 eV and the incident electron energy ($E_i$) is
equal to 100 eV. (*Closed Squares*) refer to PDDCS for H+ ion and (*closed circles*) refer to the
DDCS. The figure also shows the published experimental values of Shyn et al. (1981) (denoted by
*multiplication symbols*) and DuBois and Rudd (1978) (denote by *open circles*) and the results of
Rudd (1991) denoted by *line*) and Shultz et al. (1992) (denoted by *dotted line*)

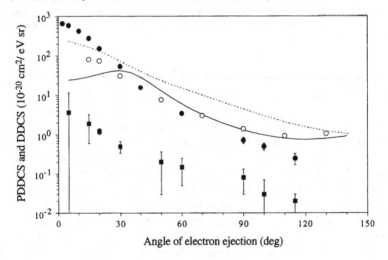

**Fig. 5.30** Present measurements of partial doubly differential cross sections (PDDCS) for ionization and doubly differential cross sections (DDCS) as a function of the angle of electron ejection
($\theta_\delta$). The secondary electron energy ($E_\delta$) is equal to 60 eV and the incident electron energy ($E_i$) is
equal to 100 eV. (*Closed squares*) refer to PDDCS for H+ ion and (*closed circles*) refer to the
DDCS. The figure also shows the published experimental values of Shyn et al. (1981) denoted by
*multiplication symbols*) and DuBois and Rudd (1978) (denoted by *open circles*) and the results of
Rudd (1991) (denoted by *line*) and Shultz et al. (1992) (denoted by *dotted line*)

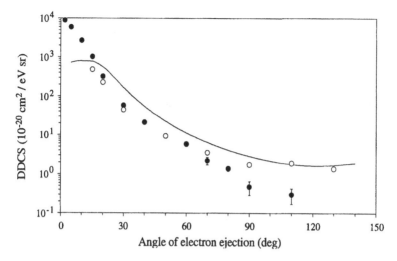

**Fig. 5.31** Present measurements of partial doubly differential cross sections (PDDCS) for ionization and doubly differential cross sections (DDCS) as a function of the angle of electron ejection ($\theta_\delta$). The secondary electron energy ($E_\delta$) is equal to 80 eV and the incident electron energy ($E_i$) is equal to 100 eV. (*Closed squares*) refer to PDDCS for H+ ion and (*closed circles*) refer to the DDCS. The figure also shows the published experimental values of Shyn et al. (1981) (denoted by *multiplication symbols*) and DuBois and Rudd (1978) (denoted by *open circles*) and the results of Rudd (1991) (denoted by *dashed line*)

present measurements agree quite well with those of DuBois and Rudd (1978) and the theoretical values of Rudd (1991) almost for the entire range but they differ from those of Shyn et al. (1981) for $\theta_\delta < 30°$ in that their data has a sharp rise at lower ejection angles. Figure 5.29 shows a good agreement between the present results and those of Shyn et al. (1981) for the whole range of values for $E_\delta = 40$ eV. Figure 5.29 also shows better agreement with the theoretical values of Rudd (1991) for $\theta_\delta \geq 40°$ and with Shultz et al. (1992) for $\theta_\delta \leq 30°$. The difference between the present results and those of DuBois and Rudd (1978) shown in Fig. 5.30, which give the results of the measurements at a detected electron energy of 60 eV, is quite large. In this case, much stronger scattering in the forward direction and a more rapid decrease in the DDCS with increasing angle is indicated by the present results. A similar trend is indicated in Fig. 5.31 which shows the results of the measurement for the detected electron energy of 80 eV.

An interesting feature of the present results is a broad maximum which occurs in the measurements of DuBois and Rudd (1978) and Shyn et al. (1981), as well. Due to the law of conservation of momentum in the binary collision between an incident electron and an atomic electron, a broad maximum is, in fact, expected (Ogurtsov 1973; Kim 1983; Rudd 1991) in the cross section at an angle given approximately by the following equation:

$$E_\delta = E_i \cos^2\theta_\delta - I,$$

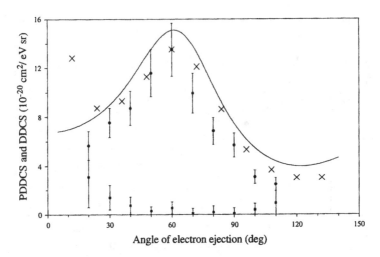

**Fig. 5.32** Present measurements of partial doubly differential cross sections (PDDCS) for ionization and doubly differential cross sections (DDCS) as a function of the angle of electron ejection ($\theta_\delta$). The secondary electron energy ($E_\delta$) is equal to 20 eV and the incident electron energy ($E_i$) is equal to 150 eV. (*Closed squares*) refer to PDDCS for H+ ion and (*closed circles*) refer to the DDCS. The figure also shows the published experimental values of Shyn et al. (1981) (denoted by *multiplication symbols*) and DuBois and Rudd (1978) (denoted by *open circles*) and the results of Rudd (1991) (denoted by *line*)

where $I$ in eV is the binding energy of the molecular electrons. Using the value of 15.43 eV from Shyn et al. (1981) for $E_i = 100$ eV and $E_\delta = 20$ eV the maximum is expected at $\theta_\delta = 53.5°$, whereas for $E_i = 100$ eV and $E_\delta = 30$ eV, the maximum is expected at 47.6°. In Figs. 5.27 and 5.28, corresponding to the measurements with $E_i = 100$ eV and $E_\delta = 20$ and 30 eV, respectively, the maxima occur at angles $\theta_\delta = 60 \pm 5°$ and $50 \pm 5°$ in reasonable agreement with the above predictions. The above equation and the present results also show that the maximum in Figs. 5.27 and 5.28 shifts to lower angles as the energy of the detected electron increases so that eventually the maximum becomes obscured by the strong increase in the electron scattering in the forward direction for larger values of detected electron energy.

Figure 5.33 summarizes the present results of DDCS as a function of $E_\delta$ for different angles of electron detection. The measured values of DDCS are relatively higher for lower $\theta_\delta$ and higher $E_\delta$ values. This result is quite understandable since electrons having higher energies are known to be preferentially scattered in the forward direction.

## 5.2.2 Sulphur Dioxide

An electron impact can produce a direct molecular ionization of $SO_2$ molecule, that is

$$e + SO_2 \rightarrow SO_2^+ + 2e,$$

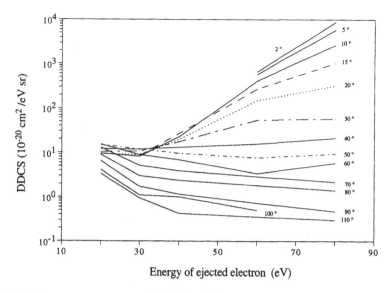

**Fig. 5.33** Present measurements of DDCS for ionization of hydrogen as a function of the secondary electron energy ($E_\delta$) for the angle of electron ejection $\theta_\delta$ having values from 2 to 110°

besides several channels of dissociative ionization (Reese et al. 1958; Smith and Stevenson 1981; Cadez et al. 1983; Curtis and Eland 1985; Eland et al. 1986; Dujardin et al. 1987; Cooper et al. 1991; Burgt et al. 1992; Masouka, 1993) such as the following:

$$e + SO_2 \rightarrow SO^+ + O + 2e$$

$$\rightarrow SO^+ + O^+ + 3e$$

$$\rightarrow SO^+ + O^- + e$$

$$\rightarrow SO + O^+ + 2e$$

$$\rightarrow SO^{++} + O + 3e$$

$$\rightarrow S^+ + O^+ + O + 2e$$

$$\rightarrow S + O^+ + O + 2e$$

$$\rightarrow S + O_2 + 2e$$

$$\rightarrow S^+ + 2O + 2e$$

$$\rightarrow S_2^{++*} + 3e \rightarrow SO^+ + O^+ + 3e$$

$$\rightarrow SO_2^{+++*} + 3e \rightarrow S^+ + O_2 + 3e$$

$$\rightarrow SO_2^{+++*} + 3e \rightarrow S^+ + O^+ + O + 3e$$

$$\rightarrow SO_2^{+++*} + 3e \rightarrow S + O^+ + O^+ + O^+ + 3e$$

$$\rightarrow SO_2^{++++*} + 4e \rightarrow S^+ + O^+ + O^+ + 4e$$

where ions such as $SO_2^+$, $SO^+$ and $(S^+, O_2^+)$ are mainly detected for the electron energies used in the present investigations. Other ions like $O^+$ and $SO^{++}$ were also observed at much lower intensities.

Figure 5.34 shows a time of flight (TOF) spectrum for the ionization of $SO_2$ molecule under electron impact where ions such as $SO_2^+$, $SO^+$, $(S^+, O_2^+)$, $O^+$ and $SO^{++}$ have been observed. The ions $S^+$ and $O_2^+$ appear at the same mass over charge ratio of 32 in this spectrum and cannot be separated in such an ion analyzer. Furthermore, $SO_2^{++}$ ion can also contribute to the build-up of the spectrum in the same position as these ions. However, $SO_2^{++}$ ion is a metastable ion (Reese et al. 1958; Cooks et al. 1974; Smith and Stevenson 1981; Dujardin et al. 1984) and dissociates to form one or more fragment ions. There is some evidence (Smyth and Muller 1933; Reese et al. 1958 and Orient and Srivastava 1984) that the peak at mass over charge ratio of 32 refers mostly to the $S^+$ ions.

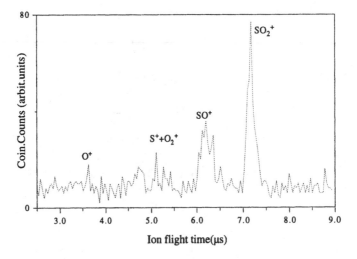

**Fig. 5.34** Time of flight (TOF) spectrum for $SO_2$ ions detected in coincidence with the electrons scattered with an energy of 432 eV and making an angle of 45° with the direction of the incident electron beam. The incident electron energy is 500 eV

The present measurements of partial doubly differential cross sections (PDDCS) for the ionization of $SO_2$ molecule are for incident electron energies ($E_i$) of 100, 150 and 500 eV and for detected electron energies ($E_\delta$) between 18–85 eV, 18–126 eV and 287–484 eV, respectively. Measurements have also been made for the angular variations of PDDCS for different ions for $E_\delta$ equal to 20 eV and $E_i$ equal to 100 and 500 eV. These data have been transformed into percentage branching ratios (%BR) and compared with the available data. The only data available in the literature for the ionization of $SO_2$ molecule are for the singly (non-differential) partial and total ionization cross sections and branching ratios by electron and photon impact.

Figures 5.35–5.43 show the present values of PDDCS for the production of $SO_2^+$, $SO^+$, ($S^+$, $O_2^+$), $O^+$ and $SO^{++}$ as a function of the detected electron energy ($E_\delta$) for incident electron energies ($E_i$) of 100, 150 and 500 eV when the angle of ejection of the detected electron $\theta_\delta$ is equal to 45°. Figures 5.37, 5.40 and 5.43 also show the values of DDCS ($=\sum$PDDCS) as a function of $E_\delta$ for incident electron energies of 100, 150 and 500 eV, respectively, when the angle of ejection of the detected electron is 45°. The curves, for the PDDCS for $SO_2^+$, $SO^+$ and ($S^+$, $O_2^+$) ions in the Figs. 5.35–5.43, in general, show peaks at different ejected electron energies. The most prominent peak is for the singly ionized molecular ion $SO_2^+$. The maxima in the DDCS occur at $E$ ($=E_i - E_\delta$) of approximately 28, 33 and 86 eV and for $E_i$ equal to 100, 150 and 500 eV, respectively. For incident electron energies of 100 and 150 eV, PDDCS for almost all ions show an increase in their values for an increase in the value of $E$ after wide minima which occur in the regions

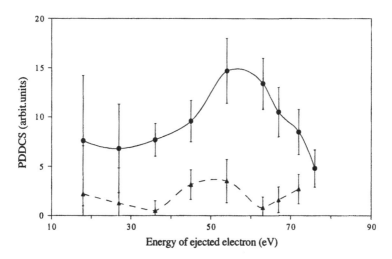

**Fig. 5.35** The present measurements of PDDCS for $O^+$ and ($S^+$, $O_2^+$) ions as a function of the ejected electron energy ($E_\delta$) when the incident electron energy ($E_i$) is equal to 100 eV and the angle of electron ejection ($\theta_\delta$) relative to the incident electron direction is 45°. (*Closed triangles*) refer to $O^+$ and (*closed circles*) refer to ($S^+$, $O_2^+$) ions. The lines are to guide the eye

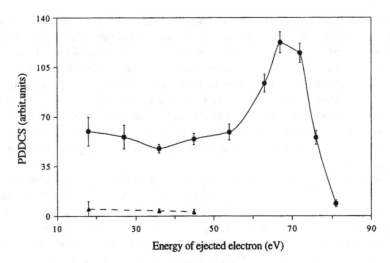

**Fig. 5.36** The present measurements of PDDCS for $SO^{++}$ and $SO^+$ ions as a function of the ejected electron energy ($E_\delta$) when the incident electron energy ($E_i$) is equal to 100 eV and the angle of electron ejection ($\theta_\delta$) relative to the incident electron direction is 45°. (*Closed triangles*) refer to $SO^{++}$ and (*closed circles*) refer to $SO^+$ ions. The lines are to guide the eye

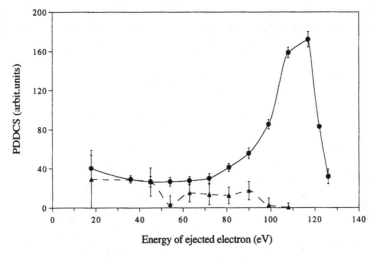

**Fig. 5.37** The present measurements of PDDCS for $SO_2^+$ ions and the DDCS for ionization as a function of the ejected electron energy ($E_\delta$) when the incident electron energy ($E_i$) is equal to 100 eV and the angle of electron ejection ($\theta_\delta$) relative to the incident electron direction is 45°. (*Closed triangles*) refer to $SO_2^+$ ions and (*closed circles*) refer to the DDCS. The lines are to guide the eye

30–40 eV and 40–70 eV for the incident electron energies of 100 and 150 eV, respectively. The value of PDDCS for each ion approaches zero as $E$ approaches the appearance potential (AP) for that ion.

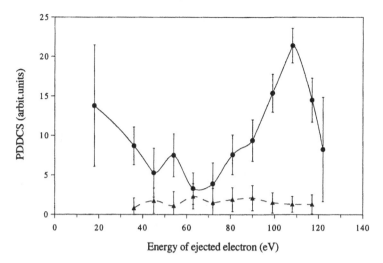

**Fig. 5.38** The present measurements of PDDCS for $O^+$ and $(S+, O_2^+)$ ions as a function of the ejected electron energy ($E_\delta$) when the incident electron energy ($E_i$) is equal to 150 eV and the angle of electron ejection ($\theta_\delta$) relative to the incident electron direction is 45°. (*Closed triangles*) refer to $O^+$ and (*closed circles*) refer to $(S+, O_2^+)$ ions. The lines are to guide the eye

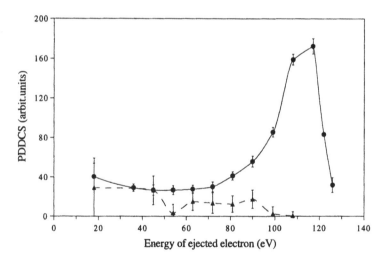

**Fig. 5.39** The present measurements of PDDCS for $SO^{++}$ and $SO^+$ ions as a function of the ejected electron energy ($E_\delta$) when the incident electron energy ($E_i$) is equal to 150 eV and the angle of electron ejection ($\theta_\delta$) relative to the incident electron direction is 45°. (*Closed triangles*) refer to $SO^{++}$ and (*closed circles*) refer to $SO^+$ ions. The PDDCS for $SO^{++}$ has been multiplied by 5 for better presentation. The lines are to guide the eye

Figures 5.44–5.49 show the present measurements of PDDCS for the production of $SO_2^+$, $SO^+$, $(S^+, O_2^+)$, $O^+$ and $SO^{++}$ ions as a function of the angle of the detected electron ($\theta_\delta$) for the detected electron energy ($E_\delta$) equal to 20 eV and

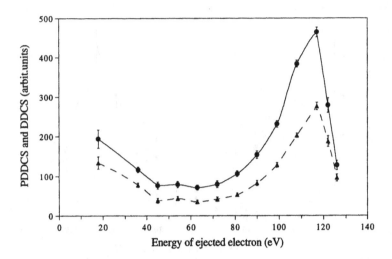

**Fig. 5.40** The present measurements of PDDCS for $SO_2^+$ ions and the DDCS for ionization as a function of the ejected electron energy ($E_\delta$) when the incident electron energy ($E_i$) is equal to 150 eV and the angle of electron ejection ($\theta_\delta$) relative to the incident electron direction is 45°. (*Closed triangles*) refer to $SO_2^+$ ions and (*closed circles*) refer to the DDCS. The lines are to guide the eye

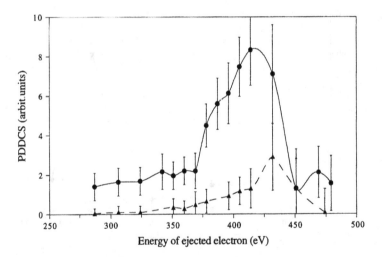

**Fig. 5.41** The present measurements of PDDCS for $O^+$ and ($S^+$, $O_2^+$) ions as a function of the ejected electron energy ($E_\delta$) when the incident electron energy ($E_i$) is equal to 500 eV and the angle of electron ejection ($\theta_\delta$) relative to the incident electron direction is 45°. (*Closed triangles*) refer to $O^+$ and (*closed circles*) refer to ($S^+$, $O_2^+$) ions. The lines are to guide the eye

incident electron energies ($E_i$) of 100 and 500 eV. Figures 5.46 and 5.49 also show the values of DDCS as a function of the angle of the detected electron ($\theta_\delta$) for the detected electron energy ($E_\delta$) equal to 20 eV and incident electron energies ($E_i$) of

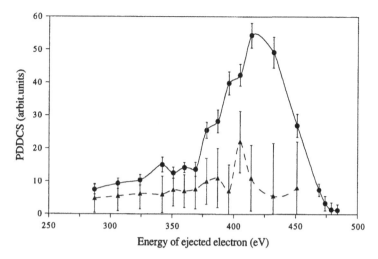

**Fig. 5.42** The present measurements of PDDCS for $SO^{++}$ and $SO^+$ ions as a function of the ejected electron energy $(E_\delta)$ when the incident electron energy $(E_i)$ is equal to 500 eV and the angle of electron ejection $(\theta_\delta)$ relative to the incident electron direction is 45°. (*Closed triangles*) refer to $SO^{++}$ and (*closed circles*) refer to $SO^+$ ions. The PDDCS for $SO^{++}$ has been multiplied by 10 for a better presentation. The lines are to guide the eye

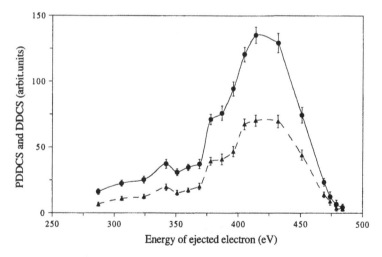

**Fig. 5.43** The present measurements of PDDCS for $SO_2^+$ ions and the DDCS for ionization as a function of the ejected electron energy $(E_\delta)$ when the incident electron energy $(E_i)$ is equal to 500 eV and the angle of electron ejection $(\theta_\delta)$ relative to the incident electron direction is 45°. (*Closed triangles*) refer to $SO_2^+$ ions and (*closed circles*) refer to the DDCS. The lines are to guide the eye

100 and 500 eV. All the curves for PDDCS and DDCS in Figs. 5.44–5.49 show a bias towards the forward direction except for $SO^{++}$ ion which remains almost constant for the whole range within the limits of error.

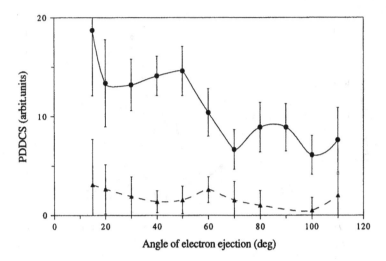

**Fig. 5.44** The present measurements of PDDCS for $O^+$ and $(S^+, O_2^+)$ ions as a function of the angle of electron ejection ($\theta_\delta$) relative to the incident electron direction when the ejected electron energy ($E_\delta$) is equal to 20 eV and the incident electron energy ($E_i$) is equal to 100 eV. (*Closed triangles*) refer to $O^+$ and (*closed circles*) refer to $(S^+, O_2^+)$ ions. The lines are to guide the eye

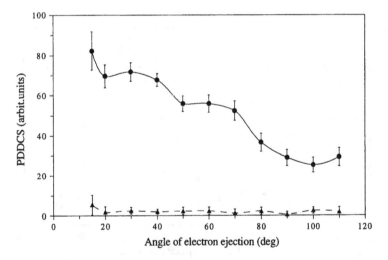

**Fig. 5.45** The present measurements of PDDCS for $SO^{++}$ and $SO^+$ ions as a function of the angle of electron ejection ($\theta_\delta$) relative to the incident electron direction when the ejected electron energy ($E_\delta$) is equal to 20 eV and the incident electron energy ($E_i$) is equal to 500 eV. (*Closed triangles*) refer to $SO^{++}$ and (*closed circles*) refer to $SO^+$ ions. The lines are to guide the eye

Figures 5.50–5.54 show the present measurements of the percentage branching ratios (%BR) of PDDCS and DDCS for the production of $SO_2^+$, $SO^+$, $(S^+, O_2^+)$, $O^+$ and $SO^{++}$ ions as a function of $E$ ($= E_i - E_\delta$), the difference in the incident and

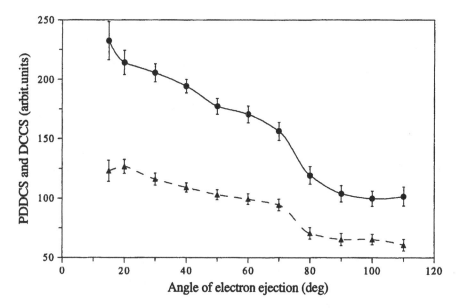

**Fig. 5.46** The present measurements of PDDCS for $SO_2^+$ ions and the DDCS for ionization as a function of the angle of electron ejection ($\theta_\delta$) relative to the incident electron direction when the ejected electron energy ($E_\delta$) is equal to 20 eV and the incident electron energy ($E_i$) is equal to 100 eV. (*Closed triangles*) refer to $O^+$ and (*closed circles*) refer to $(S^+, O_2^+)$ ions. The lines are to guide the eye

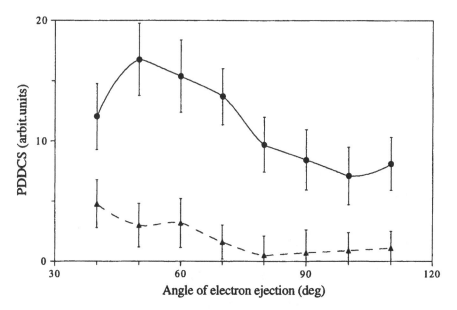

**Fig. 5.47** The present measurements of PDDCS for $O^+$ and $(S^+, O_2^+)$ ions as a function of the angle of electron ejection ($\theta_\delta$) relative to the incident electron direction when the ejected electron energy ($E_\delta$) is equal to 20 eV and the incident electron energy ($E_i$) is equal to 500 eV. (*Closed triangles*) refer to $O^+$ and (*closed circles*) refer to $(S^+, O_2^+)$ ions. The lines are to guide the eye

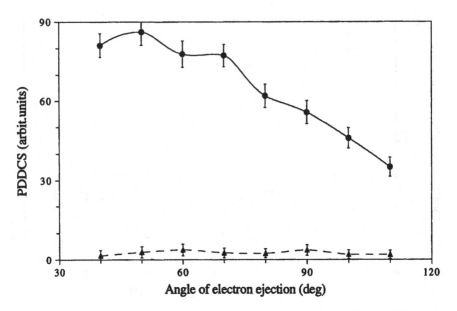

**Fig. 5.48** The present measurements of PDDCS for SO$^{++}$ and SO$^+$ ions as a function of the angle of electron ejection ($\theta_\delta$) relative to the incident electron direction when the ejected electron energy ($E_\delta$) is equal to 20 eV and the incident electron energy ($E_i$) is equal to 500 eV. (*Closed triangles*) refer to SO$^{++}$ and (*closed circles*) refer to SO$^+$ ions. The lines are to guide the eye

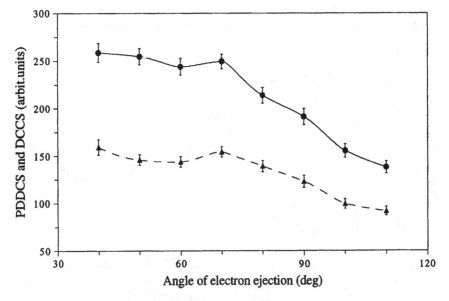

**Fig. 5.49** The present measurements of PDDCS for SO$_2^+$ and the DDCS for ionization as a function of the angle of electron ejection ($\theta_\delta$) relative to the incident electron direction when the ejected electron energy ($E_\delta$) is equal to 20 eV and the incident electron energy ($E_i$) is equal to 500 eV. (*Closed triangles*) refer to SO$^{++}$ and (*closed circles*) refer to the DDCS for ionization. The lines are to guide the eye

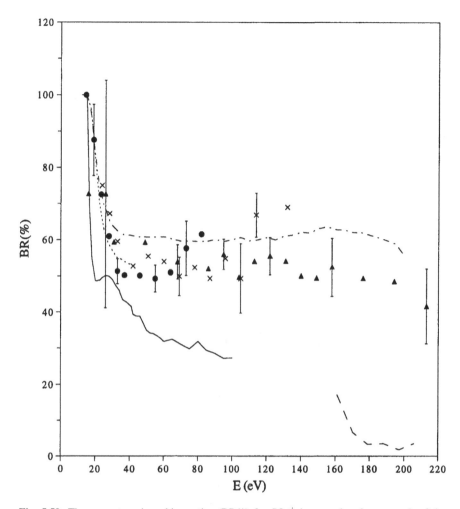

**Fig. 5.50** The percentage branching ratios (BR%) for $SO_2^+$ ions produced as a result of the electron impact ionization of $SO_2$ are plotted against $E(= E_i - E_\delta)$, where $E_i$ and $E_\delta$ are the incident and ejected electron energies, respectively. Present measurements are for 100, 150 and 500, incident electron energies. (•) refers to 100 eV, (x) refers to 150 eV and (▲) refers to 500 eV, incident electron energies. Also (—————) refers to the photoionization data of Cooper et al. (1991a, b), (........) refers to the electron impact ionization data of Smith and Stevenson (1981) and (-.-.-.-) refers to data of Orient and Srivastava (1984), where $E$ in their cases corresponds to the incident photon or electron energy

detected electron energies. In the present measurements the incident electron energies ($E_i$) of 100, 150 and 500 eV are used and $E_\delta$ energies are chosen such that $E$ has values between 15 and 213 eV. Since no published data for the branching ratios of PDDCS and DDCS for the ions produced in the electron impact ionization of $SO_2$ molecule are available, the present measurements are compared with those of branching ratios for the total ionization of $SO_2$ molecule by electron impact and

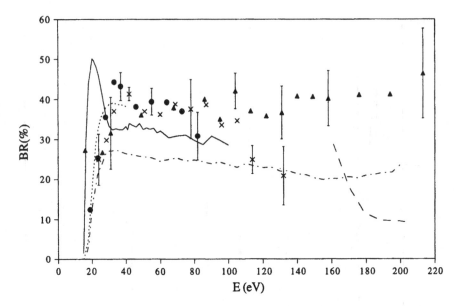

**Fig. 5.51** The percentage branching ratios (BR%) for SO$^+$ ions produced as a result of the electron impact ionization of SO$_2$ are plotted against $E(= E_i - E_\delta)$, where $E_i$ and $E_\delta$ are the incident and ejected electron energies, respectively. Present measurements are for 100, 150 and 500, incident electron energies. ($\bullet$) refers to 100 eV, (x) refers to 150 eV and ($\blacktriangle$) refers to 500 eV, incident electron energies. Also (————) refers to the photoionization data of Cooper et al. (1991a, b), (.......) refers to the electron impact ionization data of Smith and Stevenson (1981) and (-.-.-.-) refers to data of Orient and Srivastava (1984), where $E$ in their cases corresponds to the incident photon or electron energy

photoionization, where the incident electron or photon energy is equal to $E$ (as defined above).

However, this comparison is valid only if it is assumed that in the ionization of SO$_2$ molecule, ejected electrons are scattered isotropically in respect of both energy and angle of ejection and also that $E$ is approximately equal to $E_i$, that is $E_i \gg E_\delta$. Figures 5.50–5.54 also show, for comparison, the published results of Smith and Stevenson (1981) and Orient and Srivastava (1984) for percentage branching ratios for the production of different ions of SO$_2$ for low energy electron impact ionization and the results of Cooper et al. (1991) for high energy electron impact ionization with small momentum transfer or simulated photo-ionization as it is usually called. Present measurements show a general agreement with the results of Smith and Stevenson (1981) and Orient and Srivastava (1984) within their estimated error of 15%. There is, however, poor agreement with the results of Cooper et al. (1991) above $E = 30$ eV. This disagreement is more evident in the case of (S$^+$, O$_2^+$) and O$^+$ ions. A decrease in the branching ratios of the parent SO$_2^+$ ion and an increase in the case of other ions as a function of energy $E$ suggests dissociation of SO$_2$ into its constituents.

Figure 5.55 shows the present measurements of angular distribution of percentage branching ratios for SO$_2^+$, SO$^+$, (S$^+$, O$_2^+$), O$^+$ and SO$^{++}$ ions at $E_\delta = 20$ eV

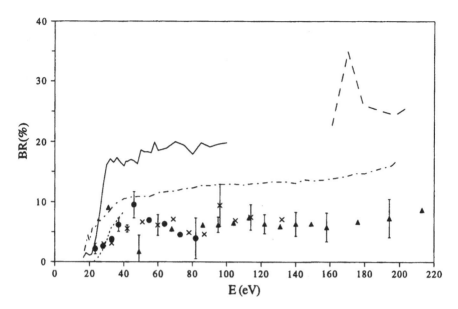

**Fig. 5.52** The percentage branching ratios (BR%) for $(S^+, O_2^+)$ ions produced as a result of the electron impact ionization of $SO_2$ are plotted against $E(= E_i - E_\delta)$, where $E_i$ and $E_\delta$ are the incident and ejected electron energies, respectively. Present measurements are for 100, 150 and 500 eV, incident electron energies. (•) refers to 100 eV, (x) refers to 150 eV and (▲) refers to 500 eV, incident electron energies. Also (————) refers to the photoionization data of Cooper et al. (1991a, b), (.......) refers to the electron impact ionization data of Smith and Stevenson (1981) and (-.-.-.-) refers to data of Orient and Srivastava (1984), where $E$ in their cases corresponds to the incident photon or electron energy

and $E_i$ equal to 100 and 500 eV. It can be said that, within the limits of error, the branching ratios for the production of different ions are independent of the angle of electron ejection. There is, however, a slight indication of increase in the branching ratios towards backward direction in the case of $SO_2^+$ ions. This may be due to the repulsive post collision interaction (PCI) with the forwardly scattered faster projectile electrons. For the 20 eV detected electrons associated with the $SO^+$ ions, the backward biasing due to PCI does not seem to be present due perhaps to the dissociative nature of the interaction which can produce more than one electron, thereby complicating the effect of PCI. In fact, there appears to be a slight forward bias giving some evidence that the detected electrons could be identified as the projectile electrons at least in the case of $E_i = 100$ eV.

### 5.2.3 Sulphur Hexaflouride ($SF_6$)

Measurements are carried out of the PDDCS for $SF_x^+$ ($x = 0$–5) and $SF_x^{++}$ ($x = 2$–5) ions produced from the $SF_6$ molecule by the impact of electrons with energies

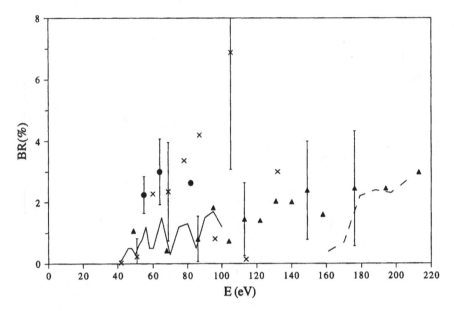

**Fig. 5.53** The percentage branching ratios (BR%) for SO$^{++}$ ions produced as a result of the electron impact ionization of $SO_2$ are plotted against $E(= E_i - E_\delta)$, where $E_i$ and $E_\delta$ are the incident and ejected electron energies, respectively. Present measurements are for 100, 150 and 500 eV, incident electron energies. (•) refers to 100 eV, (x) refers to 150 eV and (▲) refers to 500 eV, incident electron energies. Also (————————) refers to the photoionization data of Cooper et al. (1991a, b), (........) refers to the electron impact ionization data of Smith and Stevenson (1981) and (-.-.-.-) refers to data of Orient and Srivastava (1984), where $E$ in their cases corresponds to the incident photon or electron energy

($E_i$) of 100 and 200 eV. In general, the signal from all ions could be seen as shown in Figs. 5.56–5.58, but in many cases the ions SF$^+$ and SF$_4$$^{++}$ are not resolved at all so that in presenting the results the signal from these two ions is combined together.

Interestingly, the parent ion (SF$_6$$^+$) did not appear in any spectrum. It is believed that this ion is unstable in its symmetrical configuration and exhibits Jahn–Teller instability (Harland and Thynne 1969; Delwiche 1969; Pullen and Stockdale 1976; Hitchcock et al. 1978) with a considerable weakening of the SF$_5$$^+$–F bond leading to the dissociation to an SF$_5$$^+$ ion and an F atom. Nevertheless some authors (Delwiche 1969; Pullen and Stockdale 1976; Stanski and Adamczyk 1983) have reported evidence for the existence of the SF$_6$$^+$ ion.

Figures 5.56–5.58 also show that the relative intensities of singly charged ions (SF$_x$$^+$) for odd values of $x$ are higher than those with even values of $x$. The doubly charged ions (SF$_x$$^{++}$), on the other hand, show higher abundances for even values of $x$ than for the odd values of $x$. Hitchcock and Van der Wiel (1979) have previously noticed such behaviour and they have given an explanation based on a valence-bond description of the ion fragments. In this explanation, the singly charged ions are, considered to be, constructed from an S$^+$ ion in the valence configuration $3s^2 \, 3p^3$ together with an appropriate number of F atoms. The

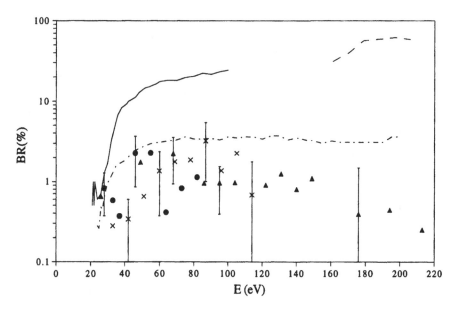

**Fig. 5.54** The percentage branching ratios (BR%) for $O^+$ ions produced as a result of the electron impact ionization of $SO_2$ are plotted against $E(=E_i - E_\delta)$, where $E_i$ and $E_\delta$ are the incident and ejected electron energies, respectively. Present measurements are for 100, 150 and 500 eV, incident electron energies. (•) refers to 100 eV, (x) refers to 150 eV and (▲) refers to 500 eV, incident electron energies. Also (————) refers to the photoionization data of Cooper et al. (1991a, b), (.......) refers to the electron impact ionization data of Smith and Stevenson (1981) and (-.-.-.-) refers to data of Orient and Srivastava (1984), where E in their cases corresponds to the incident photon or electron energy

ground-state configuration of $S^+$ implies preferential bonding to three F atoms thus explaining the higher possibility of forming the $SF_3^+$ ion. Hybridization to $3s^3$ $p^3$ $3d$ gives an $S^+$ configuration which is optimal for binding five F atoms, while no configuration can be derived to optimize bonding for $SF_4^+$ or $SF_2^+$. These ions are thus expected to be much less stable. Similarly, the valence configuration $3s^2$ $3p^2$ of $S^{++}$ implies a preferential bonding to two F atoms while $3p^3$ hybridization yields optimal bonding for $SF_4^{++}$. No such favourable scheme can be devised for $SF_x^{++}$ ions for odd $x$ values. Dibeler and Mohler (1948) also mention that the most probable dissociation process for the $SF_6^+$ ion is the loss of one fluorine atom from the $SF_6$ molecule thus helping in the formation of $SF_5^+$ ion. Further dissociation of $SF_5^+$ ion could then take place by the loss of more fluorine atoms. Frasiniski et al. (1986) in their photoioniztion experiments with $SF_6$ molecule observed similar intensity behaviour of the dissociative ions.

Figures 5.59–5.61 show the present measurements of PDDCS and DDCS for the formation of different ions from the $SF_6$ molecule by electron impact with electron incident energies ($E_i$) equal to 100 and 200 eV. Most of the data show a forward biasing especially at higher detected electron energies. This forward biasing can be interpreted as corresponding to a situation where a large momentum is transferred

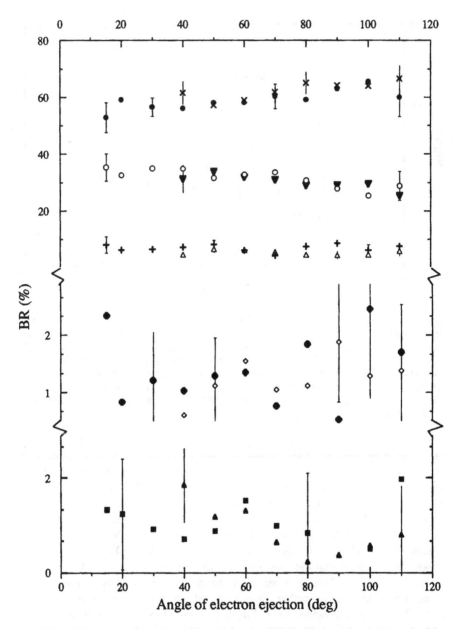

**Fig. 5.55** The present measurements of branching ratios (BR%) of ions produced as a result of the electron impact ionization of sulphur dioxide for different incident electron energies when the ejected electron energy ($E_\delta$) is equal to 20 eV. For the incident electron energy ($E_i$) equal to 100 eV, (*closed circles*) refer to $SO_2^+$, (*open circles*) refer to $SO^+$, (*plus symbols*) refer to ($S^+$,$O2^+$), (*closed diamonds*) refer to $SO^{++}$ and (*closed squares*) refer to $O^+$ ions. For the incident electron energy of 500 eV, (*multiplication symbols*) refer to $SO_2^+$, (*inverted closed triangles*) refer to $SO^+$, (*open triangles*) refer to $S^+$, $O_2^+$, (*open diamonds*) refer to $SO^{++}$ and (*closed triangles*) refer to $O^+$ ions

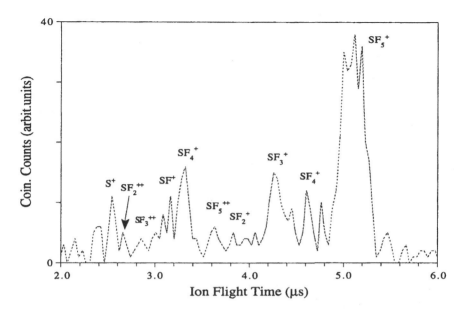

**Fig. 5.56** Time of flight (TOF) spectrum for SF6 ions as a result of the electron impact ionization for incident electron energy ($E_i$) of 100 eV, ejected electron energy ($E_\delta$) equal to 50 eV and the angle of electron ejection ($\theta_\delta$) equal to 30°

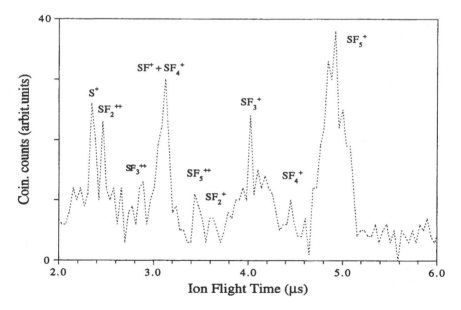

**Fig. 5.57** Time of flight (TOF) spectrum for SF6 ions as a result of the electron impact ionization for incident electron energy ($E_i$) of 200 eV, ejected electron energy ($E_\delta$) equal to 100 eV and the angle of electron ejection ($\theta_\delta$) equal to 30°

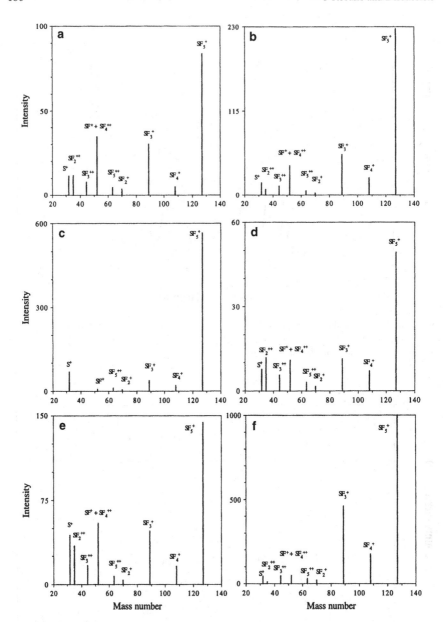

**Fig. 5.58** Mass spectra for SF6 ions as a result of the electron impact ionization. (**a**) for the incident electron energies ($E_i$) equal to 100 eV and ejected electron energies ($E_\delta$) equal to 30 eV. (**b**) For the incident electron energies ($E_i$) equal to 100 eV and ejected electron energies ($E_\delta$) equal to 50 eV. (**c**) For the incident electron energies ($E_i$) equal to 100 eV and ejected electron energies ($E_\delta$) equal to 65 eV. (**d**) For the incident electron energies ($E_i$) equal to 200 eV and ejected electron energies ($E_\delta$) equal to 50 eV. (**e**) For the incident electron energies ($E_i$) equal to 200 eV and ejected electron energies ($E_\delta$) equal to 100 eV. (**f**) For the incident electron energies ($E_i$) equal to 200 eV and ejected electron energies ($E_\delta$) equal to 150 eV

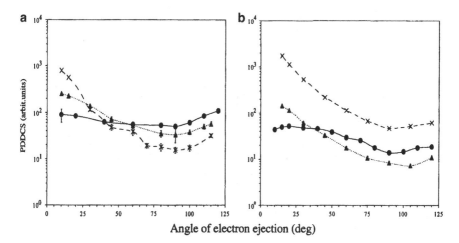

**Fig. 5.59** The present measurements of PDDCS for SF5$^+$ ion, in the electron impact ionization of SF6, as a function of the angle of electron ejection ($\theta_\delta$) relative to the incident electron direction for the incident electron energy ($E_i$) equal to: (a) 100 eV, where (*closed circles*) refer to 30 eV, (*closed triangles*) refer to 50 eV and (*multiplication symbols*) refer to 65 eV, ejected electron energies; (b) 200 eV, where (*closed circles*) refer to the ejected electron energy ($E_\delta$) of 50 eV, (*closed triangles*) refer to 100 eV and (*multiplications symbols*) refer to 150 eV

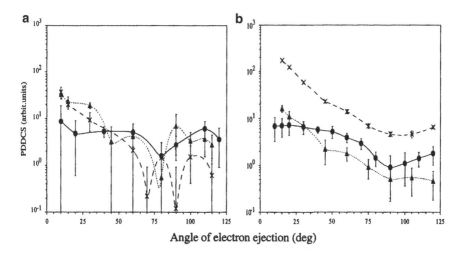

**Fig. 5.60** The present measurements of PDDCS for SF4+ ion, in the electron impact ionization of SF6, as a function of the angle of electron ejection ($\theta_\delta$) relative to the incident electron direction for the incident electron energy ($E_i$) of: (a) 100 eV, where (*closed circles*) refer to 30 eV, (*closed triangles*) refer to 50 eV and (*multiplication symbols*) refer to 65 eV, ejected electron energies; (b) 200 eV, where (*closed circles*) refer to 50 eV, (*closed triangles*) refer to 100 eV and (*multiplication symbols*) refer to 150 eV, ejected electron energies

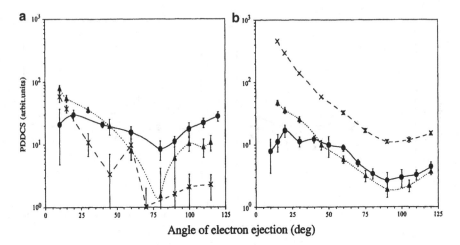

**Fig. 5.61** The present measurements of PDDCS for $SF_3^+$ ion, in the electron impact ionization of $SF6$, as a function of the angle of electron ejection ($\theta_\delta$) relative to the incident electron direction for the incident electron energy ($E_i$) of: (**a**) 100 eV, where (*closed circles*) refer to 30 eV, (*closed triangles*) refer to 50 eV and (*multiplication symbols*) refer to 65 eV, ejected electron energies; (**b**) 200 eV, where (*closed circles*) refer to 50 eV, (*closed triangles*) refer to 100 eV and (*multiplication symbols*) refer to 150 eV, ejected electron energies

from an incident electron to a target electron in the molecule or equally to a situation where the incident electron itself is scattered through a small angle with a small energy loss. Backward biasing, however, is also present particularly at lower ejected electron energies and this is due perhaps to a repulsive post-collision interaction (PCI) between slow ejected electrons and the forward-going fast scattered electrons. In the Born approximation (Vriens 1969) this biasing corresponds to collisions in which the target electron is ejected in a direction more or less opposite to the momentum, which is small in this case, transferred to it.

In the present investigations with $E_i$ equal to 100 eV and the ejected electron energy $E_\delta$ equal to 65 eV neither $SF_2^{++}$ nor $SF_2^{+++}$ ions have been observed. The absence of $SF_2^{++}$ at these energies is expected as the appearance potential energy for this ion has been measured to be 46.5 eV by electron impact (Dibeler and Mohler 1948) and 40 eV by photon impact (Hitchcock and Van der Wiel 1979). The absence of $SF_3^{++}$ at these energies suggests, as the threshold for this ion is not known at present, that the threshold appearance potential for the $SF_3^{++}$ ion must be higher than 35 eV. Interestingly, while the $SF_5^{++}$ ion has not been considered by previous investigators either because of its very low intensity or because they did not observe it at all, it has appeared at all energies used in the present measurements with a maximum occurring at $E_\delta$ equal to 65 eV and $E_i$ equal to 100 eV indicating that the appearance potential of this ion must be well below 35 eV.

Figure 5.62 shows the variation, averaged over all angles, of the percentage branching ratio (BR%), defined as the ratio of the signal produced by a particular ion to the sum of the signals from all ions, and the percentage relative abundance (RA%), defined as the ratio of a particular ion signal to the $SF_5^+$ signal, as a

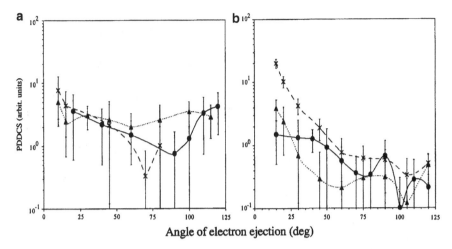

**Fig. 5.62** The present measurements of PDDCS for SF2$^+$ ion, in the electron impact ionization of SF6, as a function of the angle of electron ejection ($\theta_\delta$) relative to the incident electron direction for the incident electron energy ($E_i$) of: (**a**) 100 eV, where (*closed circles*) refer to 30 eV, (*closed triangles*) refer to 50 eV and (*multiplication symbols*) refer to 65 eV, ejected electron energies; (**b**) 200 eV, where (*closed circles*) refer to 50 eV, (*closed triangles*) refer to 100 eV and (*multiplication symbols*) refer to 150 eV, ejected electron energies

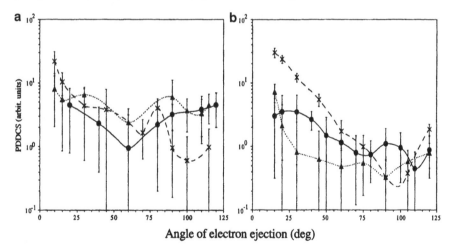

**Fig. 5.63** The present measurements of PDDCS for SF5$^{++}$ ion, in the electron impact ionization of SF6, as a function of the angle of electron ejection ($\theta_\delta$) relative to the incident electron direction for the incident electron energy ($E_i$) of: (**a**) 100 eV, where (*closed circles*) refer to 30 eV, (*closed triangles*) refer to 50 eV and (*multiplication symbols*) refer to 65 eV, ejected electron energies; (**b**) 200 eV, where (*closed circles*) refer to 50 eV, (*closed triangles*) refer to 100 eV and (*multiplication symbols*) refer to 150 eV, ejected electron energies

function of the energy loss ($E$), defined as the difference between $E_i$ and $E_\delta$ so that $E = E_i - E_\delta$. As can be seen from Fig. 5.63, for $E_i = 100$ eV, as the energy loss $E$ increases the BR% for SF$_5^+$ steadily decreases whereas the BR% and RA

% for $SF_3^+$ increase. Frees et al. (1981) have found that when $SF_5^+$ ion interacts with $SF_6$ molecule it produces $SF_3^+$ ions. However, this reaction is not very likely to happen at the low pressures used in the present experiment. For an incident energy of $E_i = 200$ eV, up to an energy loss $E = 100$ eV, the behaviour of the BR% for $SF_5^+$ and RA% of $SF_3^+$ is broadly similar to that for $E_i = 100$ eV, although the BR% of $SF_3^+$ is almost constant in this case. However, in contrast, for energy loss $E$ greater than 100 eV, the BR% of $SF_5^+$ increases with the increase in energy loss while the RA% of $SF_3^+$ decreases. Above an energy loss of 100 eV, therefore, the conversion of $SF_5^+$ to other ion states appears unlikely.

For an incident electron energy of 200 eV, Fig. 5.64 compares the angular variation of the PDDCS for $S^+$ and $SF_5^+$ for ejected electron energies $E_\delta$ of 50, 100 and 150 eV. The results for $S^+$ are typical of those obtained in this study for $S^+$, $SF_2^{++}$, $SF_3^{++}$, $SF_4^{++}$ and $SF^+$ ions whose appearance potential by electron impact lie in the range 31 to 46.5 eV. On the other hand, the results for $SF_5^+$ are typical of those obtained for the ions $SF_2^+$, $SF_3^+$, $SF_4^+$ and $SF_5^+$ whose appearance potential by electron impact lie in the range 15.7 to 26.8 eV. First of all, examining the results for $S^+$ it is seen that there is an increasing preference for scattering or ejection in the forward direction as the detected electron energy $E_\delta$ is increased and thus as the energy loss, and hence energy available to the undetected electron, decreases. Indeed with a detected electron energy $E_\delta$ of 150 eV only a maximum energy of 13 eV is available to the undetected electron. The curves for $S^+$ also suggest that the total probability of production of $S^+$, integrated over angle, decreases as $E_\delta$

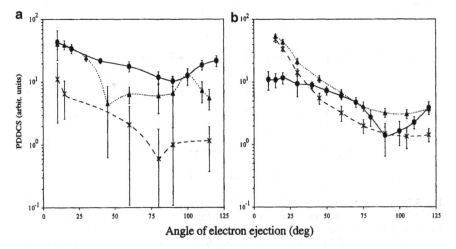

**Fig. 5.64** The present measurements of PDDCS for ($SF^+ + SF4^{++}$) ions, in the electron impact ionization of SF6, as a function of the angle of electron ejection ($\theta_\delta$) relative to the incident electron direction for the incident electron energy ($E_i$) of: (**a**) 100 eV, where (*closed circles*) refer to 30 eV, (*closed triangles*) refer to 50 eV and (*multiplication symbols*) refer to 65 eV, ejected electron energies; (**b**) 200 eV where (*closed circles*) refer to 50 eV, (*closed triangles*) refer to 100 eV and (*multiplication symbols*) refer to 150 eV, ejected electron energies

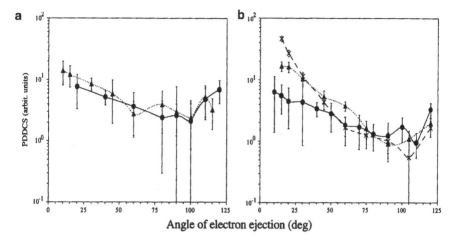

**Fig. 5.65** The present measurements of PDDCS for SF3$^{++}$ ion, in the electron impact ionization of SF6, as a function of the angle of electron ejection ($\theta_\delta$) relative to the incident electron direction for the incident electron energy ($E_i$) of: (**a**) 100 eV, where (*closed circles*) refer to 30 eV, (*closed triangles*) refer to 50 eV and (*multiplications symbols*) refer to 65 eV, ejected electron energies; (**b**) 200 eV where (*closed circles*) refer to 50 eV, (*closed triangles*) refer to 100 eV and (*multiplications symbols*) refer to 150 eV, ejected electron energies

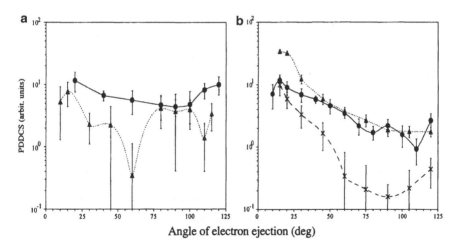

**Fig. 5.66** The present measurements of PDDCS for SF2$^{+}$ ion, in the electron impact ionization of SF6, as a function of the angle of electron ejection ($\theta_\delta$) relative to the incident electron direction for the incident electron energy ($E_i$) of: (**a**) 100 eV where (*closed circles*) refer to 30 eV, (*closed triangles*) refer to 50 eV and (*multiplication symbols*) refer to 65 eV, ejected electron energies; (**b**) 200 eV where (*closed circles*) refer to 50 eV, (*closed triangles*) refer to 100 eV and (*multiplication symbols*) refer to 150 eV, ejected electron energies

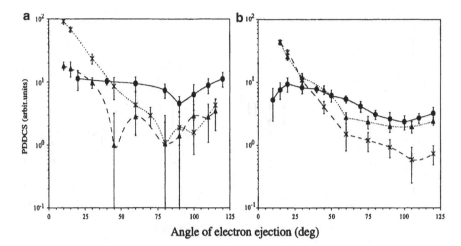

Angle of electron ejection (deg)

**Fig. 5.67** The present measurements of PDDCS for $S^+$ ion, in the electron impact ionization of SF6, as a function of the angle of electron ejection ($\theta_\delta$) relative to the incident electron direction for the incident electron energy ($E_i$) of: (**a**) 100 eV, where (*closed circles*) refer to 30 eV, (*closed triangles*) refer to 50 eV and (*multiplication symbols*) refer to 65 eV, ejected electron energies; (**b**) 200 eV, where (*closed circles*) refer to 50 eV, (*closed triangles*) refer to 100 eV and (*multiplication symbols*) refer to 150 eV, ejected electron energies

increases. This observation may be due to the fact that as the energy loss decreases fewer pathways are available for the fragmentation of the $SF_6$ molecule and the production of this particular ion (Figs. 5.65–5.67).

The results for $SF_5^+$ exhibit the same increasing preference for scattering or ejection in the forward direction as the detected electron energy $E_\delta$ is increased. However, in this case the probability of production of $SF_5^+$ integrated over angle, is much higher at a detected electron energy $E_\delta$ of 150 eV than that at 100 and 50 eV. This behaviour may result from the fact that, as the energy loss decreases and the threshold for the production of this ion is reached there is an increasing probability for the production of $SF_5^+$ and less chance of the production of other ions with higher appearance potentials.

The concluding remarks are given in the Chap. 6.

## 5.3  Measurements for the Polarization of the Fluorescent Radiation Emitted by Sodium and Potassium Atoms

### 5.3.1  Measurements for the Polarization of the Fluorescent Radiation Emitted by the Sodium Atom

Figure 5.68 shows the present measurements, of circular polarization Stokes parameter $P_3$ for the sodium resonance D-lines ($3\ ^2P_{3/2,1/2} - 3\ ^2S_{1/2}$) as a function of the applied magnetic field at the incident electron energy of 25 eV, along with the

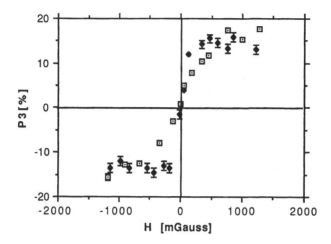

**Fig. 5.68** Present measurements of the circular polarization parameter $P_3$ for the sodium resonance lines $3\,^2P_{1/2,3/2} - 3\,^2S_{1/2}$ as a function of the applied magnetic field. The sodium atoms are polarized using a hexapole magnet (PA $\approx 21\%$). (*Open squares*) refer to the present measurement and (*closed diamonds*) refer to the data of Osimitsch (1983). Errors in our measurements are too small to be shown

measurements of Osimitch (1983). The circular polarization parameter $P_3$ is considered positive when the applied magnetic field is directed towards the observer and negative when it is directed away from the observer. The values of circular polarization parameter $P_3$ are symmetrical on the positive and negative sides of the applied magnetic field and reach a saturation value at about 0.5 Gauss. The symmetry in the values of the circular polarization $P_3$ data shows that the experimental set-up is reasonably accurate and, therefore, the measurements for linear polarization $P_1$, $P_2$ and the circular polarization parameter $P_3$ with an applied magnetic field of 0.5 Gauss can be made with some confidence.

The linear polarization of the sodium D-lines ($3\,^2P_{3/2,1/2} - 3\,^2S_{1/2}$) as a function of the incident electron energy is shown in Figs. 5.69 and 5.70 and these measurements have been made under the influence of a magnetic field of 0.5 Gauss, applied parallel to the direction of the photon detection. For comparison, along with the present measurements, are shown the linear polarization $P_1$ measurements of Osimitsch (1983), in the Fig. 5.68 and those of Jitchin et al. (1984), in the Fig. 5.70. For sodium the present measurements of linear polarization $P_1$ show a good agreement with the measurements of Osimitsch (1983) and Jitchin (1984). Figures 5.69 and 5.70 also show the various theoretical calculations of the linear polarization $P_1$. From Figs. 5.69 and 5.70 it is evident that there is a fairly good agreement between the experimental and the theoretical values of linear polarization at higher electron energy while near the threshold the theoretically predicted values are higher than the experimental values and this may, partly, be due to the energy width of the incident electron beam. The maximum value of the

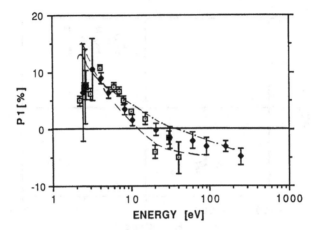

**Fig. 5.69** Present measurements of the Stokes linear polarization parameter $P_1$ of the sodium D resonance lines are shown as a function of the incident electron energy. The sodium atoms are polarized by magnetic-state selection in a hexapole magnet and are then excited by the impact of unpolarized electrons. The experimental data are not corrected for the cascade effect. ($\square$) refers to the present work and ($\blacklozenge$) refers to the work of Osimitsch (1983). Also, (————) refers to the theoretical data of Moores and Norcross (1972); (- - - -) refers to the theoretical work of Tripathi and Mathur (1973), where exchange was neglected; (-.-.-.-) refers to the data of Kennedy et al. (1977)

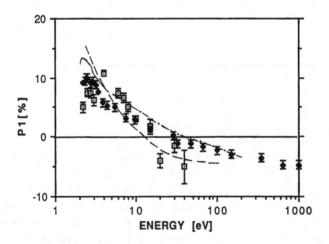

**Fig. 5.70** Present measurements of the Stokes linear polarization parameter $P_1$ of the sodium D resonance lines are shown as a function of the incident electron energy. The sodium atoms are polarized by magnetic-state selection in a hexapole magnet and are then excited by the impact of unpolarized electrons. The experimental data are not corrected for the cascade effect. ($\square$) refers to the present work and ($\blacklozenge$) refers to the work of Jitschin et al. (1984). Also, (————) refers to the theoretical data of Moores and Norcross (1972); (- - - -) refers to the theoretical work of Tripathi and Mathur (1973), where exchange was neglected; (-.-.-.-) refers to the data of Kennedy et al. (1977)

linear polarization $P_1$ occurs, however, near the threshold and is equal, approximately, to 12%.

The present experimental values of the linear polarization $P_2$ for the sodium D-lines (3 $^2P_{3/2,1/2}$ – 3 $^2S_{1/2}$), when the applied magnetic field in the interaction region is 0.5 Gauss, are shown in Fig. 5.71 along with the measurements of Osimitsch (1983). In the absence of any applied magnetic field in the interaction region the value of the linear polarization parameter $P_2$ is theoretically zero and the present measurements confirm this since even at 0.5 Gauss applied magnetic field the average value of $P_2$ is $-0.03 \pm 1.30\%$ and measurements of Osimitsch (1983) and Jitchin (1984) show the average value of $P_2$ as $0.1 \pm 1.7\%$ and $<1.5\%$, respectively. The present measurements of linear polarization parameter $P_2$ for sodium D-lines and those of Osimitch and Jitchin show, therefore, a fairly good agreement, within the experimental error, with the theoretical value of $P_2 = 0$.

The present measurements of circular polarization Stokes parameter $P_3$ for the sodium resonance D-lines (3 $^2P_{3/2,1/2}$ – 3 $^2S_{1/2}$) are shown in Figs. 5.72 and 5.73 along with the measurements of Osimitch (1983) and Jitchin (1984) and the theoretical values, for comparison. Figures 5.72 and 5.73 show that the circular polarization Stokes parameter $P_3$ increases steadily from threshold value to reach a nearly constant value above 30 eV, incident electron energy. Present measurement values of $P_3$ are in good agreement with the theoretical calculations of Moorse and Norcross (1972), and Kennedy et al. (1977), although, these are consistently larger than the previous measurements of Osimitch (1983) and Jitchin (1984). A better setting of the guiding magnetic field for the polarized atomic beam from the hexapole magnet to the interaction region, to avoid depolarization of the beam,

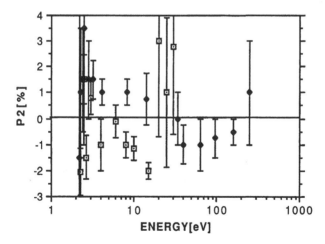

**Fig. 5.71** Present measurements of the Stokes linear polarization parameter $P_2$ of the sodium D resonance lines are shown as a function of the incident electron energy. The sodium atoms are polarized by magnetic-state selection in a hexapole magnet and are then excited by the impact of unpolarized electrons. The experimental data are not corrected for the cascade effect. (*Open squares*) refer to the present work and (*closed diamonds*) refer to the work of Osimitsch (1983). Also, (*dashed line*) refers to the theoretical data of Moores and Norcross (1972)

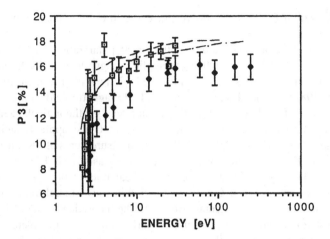

**Fig. 5.72** Present measurements of the Stokes linear polarization parameter $P_3$ of the sodium D resonance lines are shown as a function of the incident electron energy. The sodium atoms are polarized by magnetic-state selection in a hexapole magnet and are then excited by the impact of unpolarized electrons. The experimental data are not corrected for the cascade effect. ($\square$) refers to the present work and ($\blacklozenge$) refers to the work of Osimitsch (1983). The theoretical data has been calculated by Jitschin et al. (1984) and (————) refers to the theoretical work of Moores and Norcross (1972); (- - - -) refers to the theoretical work of Tripathi and Mathur (1973), where exchange was neglected; (-.-.-.-) refers to the data of Kennedy et al. (1977)

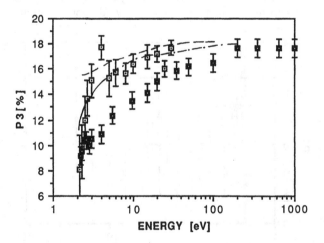

**Fig. 5.73** Present measurements of the Stokes linear polarization parameter $P_3$ of the sodium D resonance lines are shown as a function of the incident electron energy. The sodium atoms are polarized by magnetic-state selection in a hexapole magnet and are then excited by the impact of unpolarized electrons. The experimental data are not corrected for the cascade effect. ($\square$) refers to the present work and ($\blacklozenge$) refers to the work of Jitschin et al. (1984). The theoretical data has been calculated by Jitschin et al. (1984) and (————) refers to the theoretical work of Moores and Norcross (1972); (- - - -) refers to the theoretical work of Tripathi and Mathur (1973), where exchange was neglected; (-.-.-.-) refers to the data of Kennedy et al. (1977)

may be the reason for the present measurements being higher than the previous measurements of Osimitch (1983) and Jitchin (1984).

### 5.3.2 Measurements for the Polarization of the Fluorescent Radiation Emitted by the Potassium Atom

Figure 5.74 shows the present measurements of the circular polarization Stokes parameter, $P_3$, of the resonance lines ($5\,^2P_{1/2,3/2} - 3S_{1/2}$) of potassium as a function of the applied magnetic field when the incident electron energy is 25 eV. The direction of the applied magnetic field is considered to be positive along the direction of observation and negative in the opposite direction. The data for the, $P_3$, measurements in Fig. 5.74 is quite symmetrical showing that the experimental setup is reasonably accurate and, therefore, all the measurements for polarization are made with this set-up when the applied magnetic field is 0.2 Gauss.

Figures 5.75 and 5.76 show the present measurements of the linear polarization Stokes parameter, $P_1$, of potassium resonance lines ($5\,^2P_{1/2,3/2} - 3\,S_{1/2}$) as a function of the incident electron energy when the applied magnetic fields are 0.7 Gauss and 0.2 Gauss, respectively. Figure 5.75 shows the maximum value of the linear polarization, $P_1$, of about 4%, near threshold incident electron energy, which is lower than the maximum value of, $P_1$, in Fig. 5.76 of about 10%, near threshold energy, since there is a partial depolarization of the fluorescent radiation due to a higher applied magnetic field, in the interaction region, in the case of Fig. 5.75. Figure 5.76 also shows that the linear polarization Stokes parameter, $P_1$, for potassium resonance lines ($5\,^2P_{1/2,3/2} - 3S_{1/2}$), decreases in value as the incident

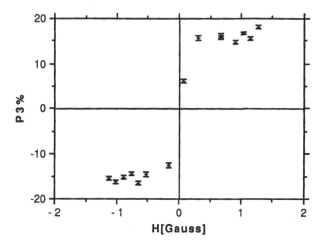

**Fig. 5.74** Present measurements of the circular polarization $P_3$ for the potassium resonance lines ($5\,^2P_{1/2,3/2} - 4\,^2S_{1/2}$) as a function of the applied magnetic field. The sodium atoms are polarized using a hexapole magnet (PA ≈ 21%). (*Open circles*) refer to the present measurement

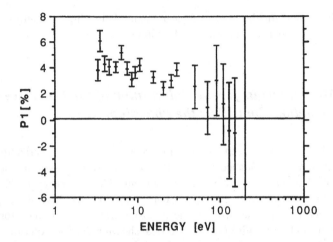

**Fig. 5.75** Present measurements of the linear polarization parameter $P_1$ potassium resonance lines $(5\,^2P_{1/2,3/2} - 4\,^2S_{1/2})$ as a function of the incident electron energy. The sodium atoms are polarized using a hexapole magnet (PA $\approx 21\%$). A magnetic field of 0.7 Gauss is applied in the interaction region

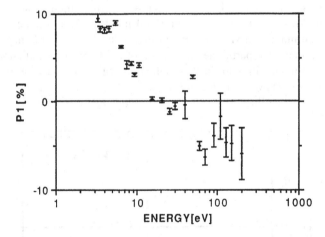

**Fig. 5.76** Present measurements of the linear polarization parameter $P_1$ potassium resonance lines $(5\,^2P_{1/2,3/2} - 4\,^2S_{1/2})$ as a function of the incident electron energy. The sodium atoms are polarized using a hexapole magnet (PA $\approx 21\%$). A magnetic field of 0.2 Gauss is applied in the interaction region

electron energy increases beyond the threshold energy and changes sign at approximately 20 eV, incident electron energy, in a great similarity to the sodium results for, $P_1$ (Figs. 5.69 and 5.70). For comparison there is no experimental or theoretical data available for potassium.

The present measurements of the linear polarization Stokes parameter, $P_2$, for the potassium resonance lines $(5\,^2P_{1/2,3/2} - 3S_{1/2})$ as a function of incident electro

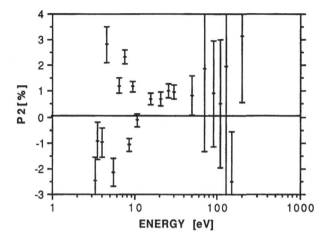

**Fig. 5.77** Present measurements of the linear polarization parameter $P_2$ of the potassium resonance lines ($5\,^2P_{1/2,3/2} - 4\,^2S_{1/2}$) as a function of the incident electron energy. The sodium atoms are polarized using a hexapole magnet (PA $\approx 21\%$). A magnetic field of 0.2 Gauss is applied in the interaction region

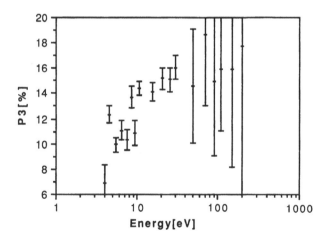

**Fig. 5.78** Present measurements of the circular polarization parameter $P_3$ of the potassium resonance lines ($5\,^2P_{1/2,3/2} - 4\,^2S_{1/2}$) as a function of the incident electron energy. The sodium atoms are polarized using a hexapole magnet (PA $\approx 21\%$). A magnetic field of 0.2 Gauss is applied in the interaction region

energy, with an applied magnetic field of 0.2 Gauss, are shown in the Fig. 5.77 and the average value of, $P_2$, is found to be 0.48 $\pm$ 1.13%.

The present measurements of the circular polarization Stokes parameter, $P_3$, for the potassium resonance lines ($5\,^2P_{1/2,3/2} - 3S_{1/2}$) as a function of incident electron energy are shown in the Fig. 5.78. Although there is no experimental or theoretical data available for comparison yet the similarity with the such measurements for sodium (Figs. 5.72 and 5.73) cannot be ignored since the general trend of $P_3$ measurements in both cases is similar and shows an increase from threshold value of 6.5% (for potassium) to nearly a constant value of 16% (for potassium). The large error bars in the values of $P_3$ for potassium are perhaps due to the generally low cross sections for the resonance lines at higher incident electron energies.

The concluding remarks are given in the Chap. 6.

## 5.4   The Excitation of Calcium and Strontium Atoms

### 5.4.1   Excitation Function of Ca II Line of Wavelength $\lambda = 393.3$ nm

The electron impact excitation function of the Ca II line having a wavelenth $\lambda = 393.3$ nm is measured from the threshold to 60 eV incident electron energy as shown in Fig. 5.79. The measurements are carried out at an oven temperature of 720 °C and with the interference filter in front of the photomultiplier tube. The signal count rate is proportional to the calcium vapour pressure and to the electron

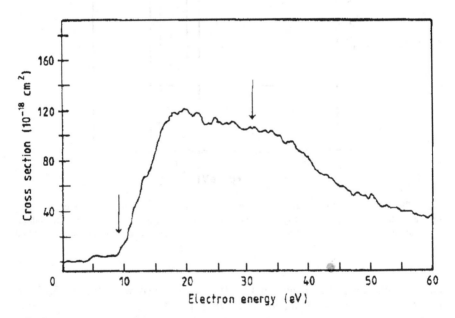

**Fig. 5.79** The direct excitation function of the Ca II $\lambda = 393.3$ nm line. The arrows near 10 and 31 eV indicate the energy thresholds of the Ca II 4 $^2$P state and the inner shell 3 p6 → 3 p5 *nl* excitation, respectively

current confirming that the single-collision process occurs dominantly. The small step near 3 eV incident electron energy and the observed intensity below the 9.25 eV threshold for the Ca II line are caused by the small fraction of light from Ca I transitions which passes through the interference filter. Above 31 e V, the threshold for the inner shell $3p^6 \rightarrow 3p^5 \, nl$ excitation which can populate the Ca II 4 $^2$P state through autoionization, a small bump is observed in the excitation function which appears to relate to this opening of another excitation channel. The small narrow structures in Fig. 5.79 are not meaningful.

## 5.4.2 Absolute Cross Section for the Ca II 4 $^2P_{1/2}$ State

The intensity ratio between the Ca I, $\lambda = 422.7$ nm and Ca II, $\lambda = 393.3$ nm lines is measured to be $22 \pm 4$ from the spectrum shown in Fig. 4.23. It is, therefore, possible to derive an absolute value for the emission cross section for the Ca II line at an electron energy of 40 eV using the absolute cross section of the Ca I, $\lambda = 422.7$ nm resonance line measured by Ehlers and Gallagher (1973). The present measurements show a value for the emission cross section of the Ca II line as $(1.02 \pm 0.19) \times 10^{-16}$ cm$^2$ which needs to be corrected for the cascade contributions, the change of the efficiency of the detecting system and the polarization of the 393.3 nm line.

The contributions from the cascade are difficult to quantify since no measurements or calculations exist for the simultaneous ionization and excitation of calcium. However, the cascade contributions can be estimated using the results of Goto et al. (1983) who studied the cross sections for the very similar electron impact excitation process of Cd atoms into low-lying Cd II states including the most important cascade fractions. Goto et al. (1983) found that at 40 eV, state of Cd II is the result of cascade from $(4d^9)^2$D states and approximately 8% of cascades from $(4d^{10})$ $^5$S and 4D states. The first group of cascade transitions do not occur in Ca since there is no full d shell. Thus we estimate the cascade contributions from the Ca II 4 $^2$P state in an analogy with the Cd II excitation to be $(14 \pm 4)$% from the 5 $^2$S and 4 $^2$D states and $(5 \pm 3)$% from the higher $^2$S and $^2$D states. Some additional population of the Ca II 4 $^2$P state can arise above 31 eV by autoionization following $3p^6 \rightarrow 3p^5 \, nl$ transitions. From the excitation function in Fig. 5.79 we estimate the contribution to be $(5 \pm 3)$% at 40 eV.

The corrections to the measured emission cross section due to the change in the efficiency of the optical components are estimated to be $-5 \pm 1$% for the grating efficiency and $-4 \pm 1$% for the quantum efficiency of the photomultiplier tube.

The measured polarization of the Ca II 393.3 nm line affects the value of the measured cross section through the anisotropy factor $(1 - P/3)$ (Percival and Seaton 1958; Forand et al. 1985) where $P$ is the degree of polarization. A polarization of $\sim$2% at 40 eV (see Fig. 5.80) leads to a correction of $-0.7 \pm 0.1$% in the value of the measured cross section. Making all these corrections, the measured value of the cross section for the Ca II 393.3 nm line is $(0.67 \pm 0.25) \times 10^{-16}$ cm$^2$.

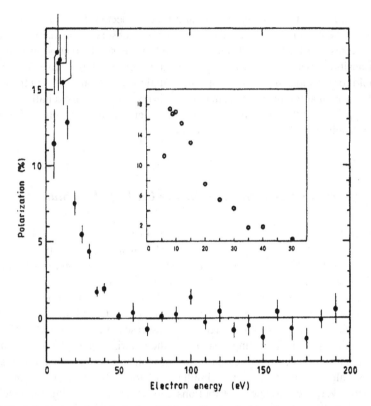

**Fig. 5.80** Polarization of the Ca II $\lambda = 393.3$ nm line excited by simultaneous electron impact ionization and excitation. The *insert* shows the polarization results in the electron energy region from threshold to 50 eV

Also 7% of the Ca II 4 $^2$P state decays to the 3 $^2$D state, therefore, making an allowance for this the absolute excitation cross section of the Ca II 4 $^2$P$_{3/2}$ state has a value of $(0.72 \pm 0.27) \times 10^{-16}$ cm$^2$ at an electron energy of 40 eV. This value is in good agreement with the measurements for Mg, Sr, Ba (summarized by Chen and Gallagher 1976) which show little variation with atomic number.

### 5.4.3  Polarization of the Ca II, $\lambda = 393.3$ nm $(4\,^2P_{3/2} \rightarrow 4\,^2S_{1/2})$ Transition

The linear polarization of the Ca II 393.3 nm line is measured as described in Sect. 4.11.1. From the profile of the filter used in the investigation and the geometrical set-up, it is estimated that approximately $20 \pm 5\%$ of the light detected by the photon counting system originates from the unpolarized light due to the

$(4\,^2P_{1/2} \rightarrow 4\,^2S_{1/2})$ transition at 396.8 nm wavelength. In addition, due to the error in the polarization measurement and any misalignment of the polarizer the necessary correction in the cross section value is judged to be $0 \pm 0.3\%$. The corrected linear polarization for the Ca II $\lambda = 393.3$ nm line is shown in Fig. 5.80 as a function of the incident electron energy. The presently measured polarization shows a maximum value of $+17\%$ near threshold and appears to converge to zero for higher electron energies. The results are in agreement with the measurements of Goto et al. (1983) for the Cd II 214.4 nm $(5\,^2P_{3/2} \rightarrow 5\,^2S_{1/2})$ line excited by single-electron impact on Cd atoms. This behaviour differs from the polarization measured for pure ion excitation by electron impact (Ca II $4\,^2S \rightarrow$ Ca II $4\,^2P$) as reported by Taylor and Dunn (1973) and from the polarization measurements for pure atomic excitation by electron impact (Ca I $4\,^1S \rightarrow$ Ca I $4\,^1P$) as reported by Ehlers and Gallagher (1973). In both cases the polarization clearly crosses from positive to negative values at the incident electron energy of approximately 40 eV.

The theory by Percival and Seaton (1958) predicts a polarization of 60% and $-42.9\%$ for a $^2P_{3/2} \rightarrow\,^2S_{1/2}$ transition at threshold and at high energy, respectively, for direct electron impact excitation of atoms. The present measurements confirm that this theory cannot be extended to simultaneous ionization and excitation nor, obviously, to ion excitation.

Bartschat and Fang (2002) have presented detailed second-order calculations of simultaneous excitation–ionization in calcium for the incident electron energy of 400 eV. Rapid and extreme variations are predicted by them in the Stokes parameters. Stevenson and Crowe (2004) have investigated the excitation–ionization of the calcium atom by electron impact for incident electron energy of 400 eV and their preliminary measurements of the linear Stokes parameters agree generally with the calculated values of Bartschat and Fang (2002).

Chapter 6 gives the concluding remarks.


### 5.4.4 The Coherence and Polarization Parameter Measurements for the Excitation the 5 $^1P$ State in Strontium

Present measurements of the Stokes parameters $P_1, P_2, P_3$ for the Sr I $(5\,^1P \rightarrow 5\,^1S)$ $\lambda = 460.7$ nm line as a function of the electron scattering angle $(\theta_e)$ for an incident electron energy of 45 eV are shown in Fig. 5.81. The error bars show the statistical uncertainty only. The uncertainties due to the finite solid angle of the photon detector and other experimental effects are found to be small. Figure 5.81 also shows the theoretical calculations of Clark et al. (1992) who used the first-order many-body theory (FOMBT) and Srivastava et al. (1992) who employed the relativistic distorted wave approximation (DWA). The present measurement values of the Stokes parameters change quite rapidly over the range of scattering angles 0–180° and this is not surprising in view of the large number of interfering partial waves affecting the scattering process for a relatively heavy atom. The present

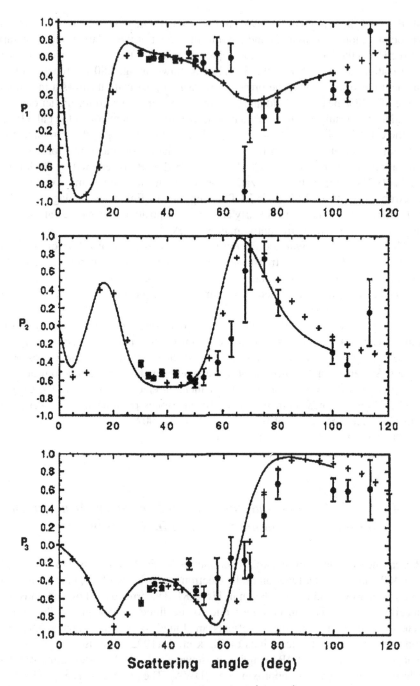

**Fig. 5.81** Stokes parameters $P_1$, $P_2$ and $P_3$ of the Sr I (5 $^1$P $\rightarrow$ 5 $^1$S) transition at $\lambda = 60.7$ nm excited by electron impact at an energy of 45 eV as a function of the electron scattering angle, *dashed line* shows the first-order many-body theory calculations by Clark and Abdallah (1992), and *plus symbols* show the distorted wave approximation calculations by Srivastava et al. (1992)

measurements show a good agreement with the theoretical calculations. However, the distorted wave calculations show a better agreement with the present measurements especially in the case of $P_3$ values.

A completely coherent excitation process manifests itself in the value of the total polarization $|P|$, which must have a numerical value equal to one, within the experimental error, for a $^1S \rightarrow {}^1P \rightarrow {}^1S$ process in light atoms with zero nuclear spin. The present measurements of Stokes polarization parameters shown in Fig. 5.81 have not been corrected for the depolarizing effect of the hyperfine structure of the odd isotope $^{87}Sr$ ($I = 9/2$, 7% abundance) which is estimated to be not more than 4% based on the hyperfine structure measurements of Kluge and Sauter (1974). Taking into account this depolarization and the experimental error bars, our measured degree of polarization $|P|$ values (shown in Fig. 5.82a) are very close to one except at large scattering angles. The reason for the lower values may be due to the convolution of the rapidly varying Stokes parameter values with the finite angular acceptance angle of the electron analyzer. This instrumental effect can distort the coherence parameters (Martus et al. 1988) and indeed the distortion has been found to be significant in the super elastic scattering of electrons from laser excited $^{138}Ba$ ($^1P_1$) atoms (Zetner et al. 1990).

The scattering parameters $\lambda$ and $\chi$ are derived from the measured Stokes parameter values and are shown in Fig. 5.82b, c). As expected from the presently measured values of Stokes parameter $P_1$, the $\lambda$ parameter shows a slow variation at smaller scattering angles and there is an indication of a sharp minimum at $\theta_e \approx 68°$. Also at $\theta_e \approx 70°$, the value of the phase parameter $\chi$ drops significantly. Such a sharp drop in the value of $\chi$ has been detected in coherence type experiments reported on the He 1 $^1S \rightarrow n\,{}^1P$ excitation (e.g. Ibraheim et al. 1985, 1986). The physical origin of this phase jump is not clear. All the measured parameters show an abrupt change at the scattering angle of $\approx 70°$. Similar change is also found in the preliminary measurements of the electron impact coherence parameters for the Sr ($5\,{}^1P$) state excitation at electron impact energies of 30.3 eV and 58.4 eV (Hamdy et al. 1992).

Figure 5.83 shows the values of the target parameters $L_\perp$, $P_1$ and $\gamma$ as a function of the scattering angle. As previously observed in $^1P$ state excitation of He (Ibraheim et al. 1986) the presently measured values (as shown in Fig. 5.83) of the angular momentum $L_\perp$, transferred to the atom normal to the scattering plane show a sign change at intermediate angles especially at 70°. At small scattering angles $L_\perp$ is positive which in the natural frame indicates the predominant population of the magnetic sublevel $m_l = +1$. Above 70° scattering angle $L_\perp$ has negative values and this indicates that the magnetic sublevel $m_l = -1$ becomes populated predominantly.

In the case of complete coherence of the excitation process a rapid change in the electron charge cloud alignment angle $\gamma$ occurs usually where the excited atom is in a nearly circular state. This behaviour of $\gamma$ can be explained with the aid of Poincaré sphere as discussed by Andersen et al. (1988). The present measurements for $\gamma$ confirm this behaviour and show a rapid change in $\gamma$ at scattering angle of 80° where the Sr ($5\,{}^1P$) state has a nearly circular shape.

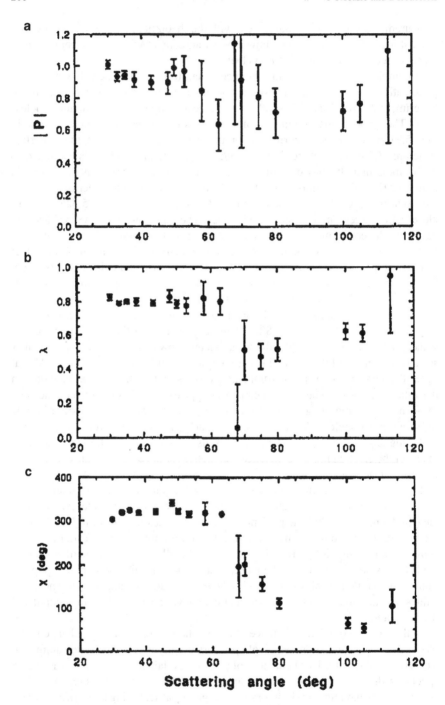

**Fig. 5.82** Results for (**a**) the degree of total polarization $|P|$, (**b**) the scattering parameter $\lambda$ and (**c**) the phase parameter $\chi$ for the Sr (5 $^1$P) state excitation with 45 eV incident electron energy

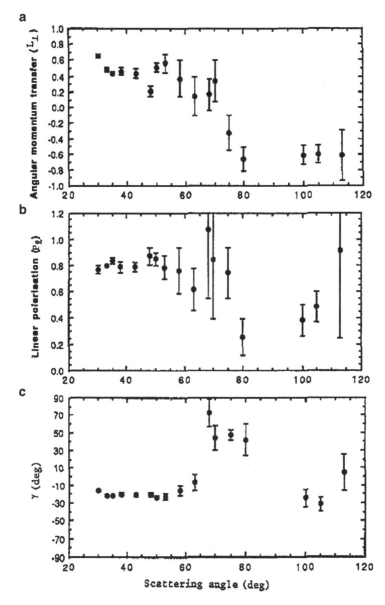

**Fig. 5.83** Results for (**a**) the angular momentum transferred $L_\perp$, (**b**) the degree of linear polarization $P_1$, and (**c**) the alignment angle of the charge cloud $\gamma$

The charge cloud is fully characterized by either of the three parameter sets $(P_1, P_2, P_3)$, $(\lambda, \chi)$ and $(L_\perp, P_1, \gamma)$. It is most directly related to $L_\perp$, $P_1$ and $\gamma$, where $L_\perp$ determines the waist of the charge cloud ($|L_\perp| = 1$ corresponds to a fully circular charge cloud) while $P_1$ determines the ratio between the width and the length of the

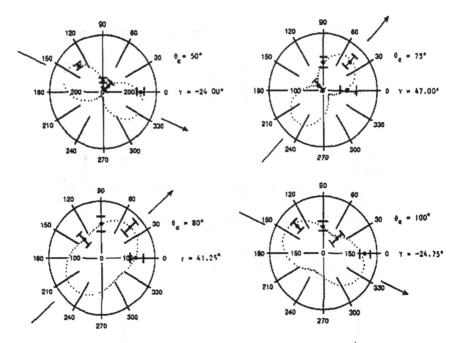

**Fig. 5.84** Shape and rotation of the electron charge cloud of the excited Sr (5 $^1$P) state for different electron scattering angles while the incident electrons enter along 0° angle. The measured points are the normalized coincidence-rates for the linear polarizer axis settings at 0, 45, 90 and 135°. The *dotted lines* are the fitted curves for the experimental data

charge cloud and $\gamma$ represents the direction of the charge cloud with respect to the incident electron direction.

The changing shape and the rotation of the p-electron charge cloud in the scattering plane are borne out by the values of $L_\perp$, $P_1$ and $\gamma$. The rapid changes in charge cloud for scattering angles between 50 and 100° are illustrated in Fig. 5.84. The measurement data points show the normalized coincidence rates at linear polarizer axis settings at 0, 45, 90 and 135° angles with respect to the incident electron direction while the curves are fitted to the experimentally measured data.

At small scattering angles, that is $\theta_e < 50°$, the excited charge cloud rotates away from the direction of the incoming electron reaching a minimum value of –24° at $\theta_e = 50°$. At larger scattering angles, the charge cloud rotates back, passing through the direction of the incoming electron beam ($\gamma = 0°$), and eventually aligns itself so that $\gamma$ becomes positive at $\theta_e > 63°$. At the electron scattering angle $\theta_e = 80°$, the charge cloud has a nearly circular shape and a rapid change in $\gamma$ occurs. Within a narrow range of scattering angles the charge cloud axis rotates back and reaches another minimum value at $\theta_e = 105°$.

Chapter 6 gives the concluding remarks.

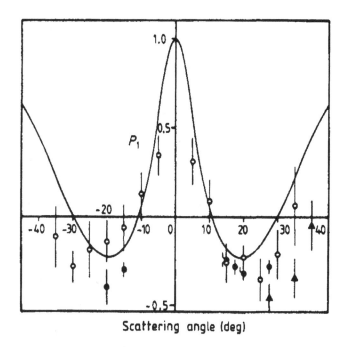

**Scattering angle (deg)**

**Fig. 5.85** Stokes parameter $P_1$ for the linear polarizer angles $\psi = 0°$, $90°$ for the 3 $^1$P state excitation of He at 80 eV for positive and negative angles. *Open circles* show the present results, *closed circles* show the measurements of Standage and Kleinpoppen (1976), *closed triangles* show the measurement values of Ibraheim et al. (1985, 1986) and *dashed line* shows the calculated values of Madison (1979, 1982)

## 5.5   Excitation of 3 $^1$P State of Helium

Figures 5.85–5.87 show the present measurements of the three Stokes parameters as a function of the scattering angle. Figure 5.85 shows the average values obtained for the Stokes parameter $P_1$, with and without the $\lambda/4$ plate. The present results in Figs. 5.85–5.87 are compared with other polarization correlation data for scattering angles $< 40°$. Contrary to the behaviour reported by Standage and Kleinpoppen (1976) it is found that $P_2$ changes sign between positive and negative scattering angles. The sign of the $P_2$ values of Standage and Kleinpoppen (1976) for negative scattering angles has been altered in Fig. 5.86, and the values are then in agreement with the present results. Although there is some spread in the measured values but most agree within the statistical uncertainties (one standard deviation is shown). It would be difficult to reduce the errors much further in view of the unfavourable branching ratios from 3 $^1$P $\rightarrow$ 2 $^1$S which result in very low signal levels and corresponding long measuring times.

The Stokes parameter measurements can be used to calculate the total polarization $P = (P_1{}^2 + P_2{}^2 + P_{\downarrow 3}^{\uparrow 2})^{1/2}$ which is expected to be equal to one as shown experimentally by Standage and Kleinpoppen (1976). While the majority of the

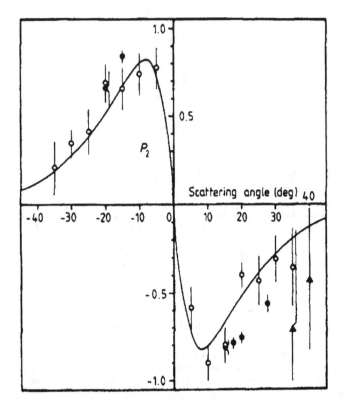

**Fig. 5.86** Stokes parameter $P_2$ for the linear polarizer angles $\psi = 45°$, $135°$ for the 3 $^1$P state excitation of He at 80 eV for positive and negative angles. *Open circles* show the present results, *closed circles* show the measurements of Standage and Kleinpoppen (1976) (it may be noted that the sign of their measurements for negative scattering angles has been altered), *closed triangles* show the measurement values of Ibraheim et al. (1985, 1986) and *dashed line* shows the calculated values of Madison (1979, 1982)

present measurement data points would not be considered inconsistent with $P = 1$, the data points collectively fall below $P = 1$ indicating that the polarization is reduced by an unaccounted systematic effect. However, this is unlikely to alter the general shape of the polarization curve in Figs. 5.85–5.87 and in any case does not affect the symmetry behaviour of the Stokes parameters.

The theoretical curves have been calculated by Madison (1979, 1982) and they represent the behaviour of the experimental results fairly well. Different theoretical models have been discussed by Beijers et al. (1987), Fon et al. (1991) and Lange et al. (2006).

Figure 5.88 shows the scattering parameters $\lambda$ and $\chi$ together with other results based on polarization correlation measurements. The present values of $\chi$ represent the average of the results obtained from Stokes parameters $P_2$ and $P_3$. The various measurements and theoretical calculations of Madison (1979) agree reasonably well. It may be noted, however, that the theoretical value of $\chi$ does not tend to

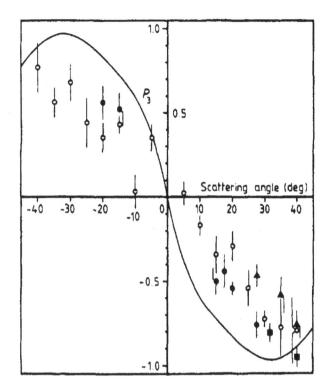

**Fig. 5.87** Stokes parameter $P_3$ for the circular polarization for the 3 $^1$P state excitation of He at 80 eV for positive and negative angles. *Open circles* show the present results, *closed circles* show the measurements of Standage and Kleinpoppen (1976), *closed triangles* shows the measurement values of Ibraheim et al. (1985, 1986), *closed squares* show the data from Beijers et al. (1986) and *dashed line* show the calculated values of Madison (1979, 1982)

zero for zero scattering angle and that there is no sudden change in the value of $\chi$ when $\theta_e$ is varied through zero.

The concluding remarks are given in Chap. 6.

# 5.6 Excitation of 3 $^3$P State of Helium

The present measurements (see Fig. 5.89) of Stokes parameters $P_1$, $P_2$ and $P_3$ are compared with the theoretical calculations of Cartwright and Csannak (1986) and show general agreement within the error bars. The negative value of the circular polarization $P_3$ proves that the orientation of the atom and thus the angular momentum transfer during the collision are positive for these scattering angles as in the case of the 3 $^1$P state (see Sect. 5.5) excitation. No change of sign of $P_3$ is found within the angular range used, in agreement with the calculations of Cartwright and Csannak (1986) which indicate a sign change at $\theta_e = 105°$. At $\theta_e = 60°$ the measured value of

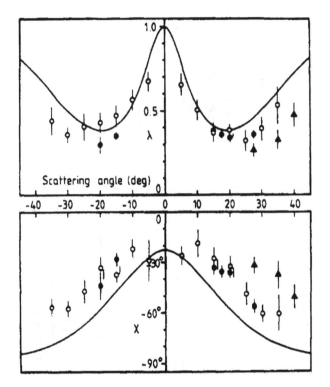

**Fig. 5.88** $\lambda$ and $\chi$ parameters for the 3 $^1$P state excitation of He at 80 eV for positive and negative angles. *Open circles* show the present results, *closed circles* show the measurements of Standage and Kleinpoppen (1976), *closed triangles* show the measurement values of Ibraheim et al. (1985, 1986) and *dashed line* shows the calculated values of Madison (1979, 1982)

$P_3$ (corrected for the depolarization) nearly reach the value $-1.0$. The calculations of Cartwright and Csannak (1986) show a minimum value of $P_3 = -1$ at $\theta_e = 70°$.

The present measurements of Stokes parameters are also used to calculate the scattering parameters $\lambda$ and $\chi$ shown in Fig. 5.90. Within the range 0 to $-2\pi$ adopted for $\chi$, $P_2$ and $P_3$ each allow two possible values of $\chi$ in line with the symmetry of the sine and cosine functions. Only two of these four results are consistant with each other and the average of these is shown in Fig. 5.90. The present values of the phase angle $\chi$ for spin-exchange excitation are not too different from those found for the 3 $^1$P state excitation (Ibraheim et al. 1986) of helium. Within the error bars the values for $\lambda$ and $\chi$ are in agreement with the theory by Cartwright and Csannak (1986).

The total polarization $P = (P_{\downarrow 1}^{\uparrow 2} + P_{\downarrow 2}^{\uparrow 2} + P_{\downarrow 3}^{\uparrow 2})^{1/2}$ and the coherence parameter

$$[\mu] = \frac{[(P_2 - iP_3)]}{(1 - P_1{}^2)^{\frac{1}{2}}}$$

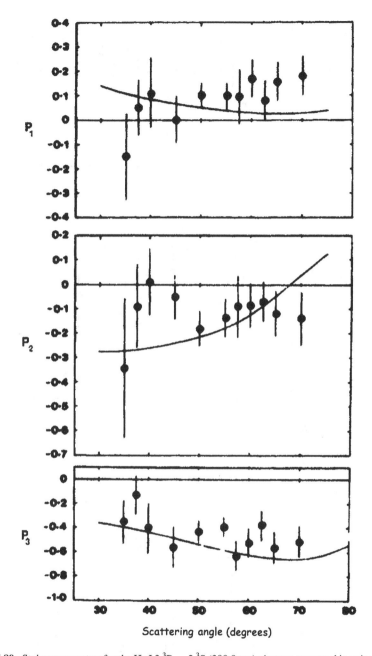

**Fig. 5.89** Stokes parameters for the He I 3 $^3$P $\rightarrow$ 2 $^3$S (388.9 nm) photons measured in coincidence with the scattered electrons following spin exchange excitation from the ground state of helium for incident electron energy of 60 eV. *Full lines* are from the theoretical results from Cartwright and Csanak (1986)

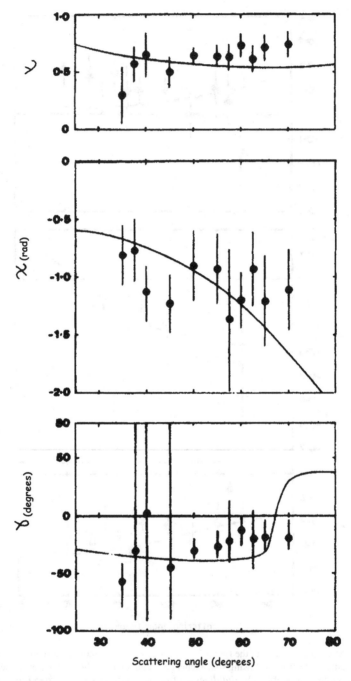

**Fig. 5.90** Scattering parameters $\lambda$, $\chi$ and $\gamma$ as a function of the scattering angle for 3 $^3$P state excitation in helium for incident electron energy of 60 eV. *Full lines* are the theoretical calculation values from Cartwright and Csanak (1986)

**Fig. 5.91** Total polarization and coherence parameters for electron-impact excitation of the 3 ³P state of helium as a function of the scattering angle when the incident electro energy is 60 eV

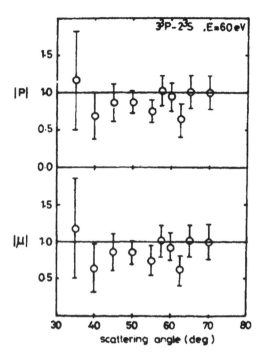

are shown in Fig. 5.91 and confirm complete polarization and coherence for the spin exchange process. Andersen et al. (1984, 1986) have used $P_3$, the degree of circular polarization, $P_l$, the degree of linear polarization, and $\gamma$ the alignment angle of the electron cloud with respect to the $z$-axis, instead of $\gamma$ and $\chi$. Here $\gamma$ is the same as $\theta_{min}$ introduced by Eminyan et al. (1973, 1974, 1975). If the total polarization can be taken to be 1, $P_3$ and $\gamma$ are sufficient to describe the excitation process completely. According to the calculations of Cartwright and Csannak (1986), who used the two-parameter approach, $\gamma$ tends to change sharply in the vicinity of scattering angles, where $[P_3]$ has a value close to 1.

Hayes and Williams (1996) have also investigated the excitation of helium 3 ³P state for incident electron energy from threshold to 300 eV using polarized electrons and they conclude that their measurements are consistent with the LS-coupling of the upper states of helium and that the non-zero value of the Stokes parameter $P_3$ indicates strong electron exchange effects particularly near threshold.

Fon et al. (1991) have used 11-state and 19-state R-matrix calculations for the excitations $1\ ^1S \rightarrow n\ ^{3,1}P$ ($n = 2$, 3 and 4) to obtain Stokes parameters and they conclude that these parameters are $n$-independent as predicted by Csanak and Cartwright (1986, 1988). Michael Lange et al. (2006) have reported calculations using R-matrix, B-spline R-matrix and the convergent close-coupling methods and have concluded that there is generally good agreement with experiments.

Chapter 6 gives the concluding remarks.

# Chapter 6
# Conclusion

**Abstract** The present analysis of the excitation and ionization of atoms and molecules by electron impact has not only provided great insight into these processes but has also made available experimental data to help build better theoretical models.

**Keywords** Atoms · Excitation · Ionization · Molecules · Polarization

The first Born approximation can satisfactorily explain single ionization at higher impact velocities compared to the orbiting electron velocity, but even in this case the processes involved in multiple ionization are not clear (Anderson et al. 1987) At lower impact velocities, however, the ionization processes become even more complicated due to the interactions between the projectile, the target nucleus and the ejected electrons, and also because of the changes in the binding energies of the target electrons due to the presence of the charged projectile (Brandt and Basbas 1983). To find appropriate approximations for the solution of the many-body problem is a challenging goal for a theoretical physicist. Accurate data are, therefore, required not only for total and partial ionization cross sections but also for differential ionization cross sections, since the latter provides more critical tests for the theoretical calculations. The present data of partial doubly differential cross section for the ionization of helium, argon, krypton and xenon will therefore be useful in extending our understanding of the phenomenon of multiple ionization and will provide a stringent test for any future theoretical model for the process.

Recently the convergent close-coupling (CCC) methodology of Bray and Stelbovics has provided the best agreement of scattering theory with experimental results (Philip and Stelbovics 2002). This method, however, is computationally intensive and is currently limited to the valence shell of atoms containing only one or two electrons. Baertschy et al. (2001) has described the exterior complex scaling (ECS) method, which requires massive parallel supercomputing to solve the three-body problem without significant approximation. This has provided very accurate

A. Chaudhry and H. Kleinpoppen, *Analysis of Excitation and Ionization of Atoms and Molecules by Electron Impact*, Springer Series on Atomic, Optical, and Plasma Physics 60, DOI 10.1007/978-1-4419-6947-7_6, © Springer Science+Business Media, LLC 2011

theoretical results for hydrogen at low incident energies, but will require significant advances in computing technology before it can be applied to larger atoms.

The results of X-ray-ion coincidence study show that highly charged xenon ions (the present value of the average charge per ion is 6.9) are associated with the emitted X-rays indicating that multiple ionization results from the vacancy cascade following the emission of an X-ray photon (Coglan and Clansing 1973) by the xenon ion which has an initial vacancy in the inner-shell. Such information can help in understanding the multiple ionization process.

Partial doubly differential cross sections (PDDCS) for the dissociative and non-dissociative ionization of $H_2$, $SO_2$ and $SF_6$ molecules by electron impact have been measured and a set of new data is made available. The present measurements of partial doubly differential cross sections (PDDCS) and the percentage branching ratios (BR(%)) for $SO_2$ molecule can add to the understanding of processes responsible for the dissociative ionization and consequent disappearance from the atmosphere of this air-polluting molecule. $SF_6$ is a much used gaseous dielectric and the possibility of its forming harmful by-products is a serious problem and the present measurements can help in understanding and overcoming these difficulties.

Some of the conclusions which can be drawn, from the investigation of the polarization of the fluorescence radiation from the electron impact excitation of spin-polarized atoms of sodium and potassium, are that the linear polarization Stokes parameter $P_2$ is zero, within the experimental error, for sodium and potassium atoms and this agrees well with the predictions of the theoretical calculations (Jitschin et al. 1984). Further that at low incident electron energy the present measurement values of the circular polarization Stokes parameter $P_3$, for sodium and potassium, are quite low ($\sim$7%) and this reveals that the three interaction channels namely direct, exchange and interference are comparable in magnitude. At higher incident electron energies the values of the circular polarization reach about 17%, for both sodium and potassium atoms, and this indicates that the contributions of the exchange process are minimal at this point.

We would like to mention that an electron-photon coincidence technique (see Chaudhry and Kleinpoppen 1993–1994) can be used to measure more accurately the Stokes polarization parameters for the resonance radiation from electron impact excitation of spin-polarized atoms of alkali metals. In this preliminary study a 30° parallel plate electrostatic analyzer is used to detect, by a channeltron, the scattered electrons with an energy loss equal to the excitation potential of the D-lines of the alkali metals. The photons of the resonance line which have passed through the polarization optics and a light filter are detected by a liquid-nitrogen-cooled photo-multiplier. Coincidences are then found between the detected electrons and photons. This technique can, in future, be used to gather more accurate information about the excitation process.

As a highly correlated process, simultaneous ionization-excitation of atoms by electron impact presents a major challenge to experimentalists and theorists alike. The work of Khiefets et al. (1999) shows that second-order effects are expected to be significantly more important in the ionization-excitation process than in direct ionization when the residual ion remains in the ground state. Although, a non-perturbative treatment of the process is, in principle, possible by using the convergent close-coupling (CCC), R-matrix with pseudo-states (RMPS), or other advanced close-coupling-type methods including a direct solution of the time-dependent Schrödinger equation, the computational effort of such attempts seems currently prohibitive (Bartschat and Fang 2002).

The electron impact excitation of helium has been a subject of great interest. The excitation of helium state ($1\,^1S - 3\,^3P$) is of special interest since it is a pure electron exchange process which can provide complete correlation information without the use of spin-polarized beams. Experimentally and theoretically it is the least studied of the $n = 3$ states using the sensitive electron-photon correlation methods (Igual-Ruiz et al. 2001).

# Bibliography

Aberg, T.: In: Craseman, B. (ed.) Atomic Inner-shell Processes, vol. 1, p. 353, Academic, New York (1975)

Admczyk, B., Boerboom, A.H.J., Lukasiewicz, M.: Int. J. Mass Specrom. Ion Phys. **9**, 407 (1972)

Afrosimov, V.V.: In: Mehlhorn, W., Brenn, R. (eds.) Proceedings of the Second Int. Conf. on Inner-Shell Ionization Phenomenon, p. 258, University of Freiberg (1976)

Aksela, H., Aksela, S.: J. Phys. B **16**, 1531 (1983)

Andersen, N., Gallagher, J.W., Hertel, I.V.: In: Lorents, D.C., Meyerhof, W.E., Peterson, J.R. (eds.) Electronic and Atomic Collision, vol. 165, pp. 57, 1–188. Elsevier, New York (1986)

Andersen, N., Gallagher, J.W., Hertel, I.V.: Phys. Rev. **165**, 1–188 (1988)

Anderson, R.J., Hughes, R.H.: Phys. Rev. A **5**, 1194–1197 (1972)

Anderson, R.J., Lee, E.T.P., Lin, C.C.: Phys. Rev. **160**, 20–22 (1967)

Auger, P.: J. Phys. Radium **6**, 205 (1925)

Avaldi, L., Camilloni, R., Multari, R., Stefani, G., Robaux, O., Tweed, R.J., Vien, G.N.: J. Phys. B **31**, 2981–2997 (1998)

Aydinol, M., Hippler, R., Mc Gregor, I.J.: Phys. Rev. B **13**, 989 (1980)

Baertschy, M., Rasciano, T.N., Mc Curdes, C.W.: Phys. Rev. A **64**, 022709 (2001)

Bambaynek, W., Crasemann, B., Fink, R.W., Freund, H.U., Mark, H., Swift, C.D., Price, R.E., Ras, P.V.: Rev. Mod. Phys. **44**, 716 (1972)

Barker, E.S.: Geophys. Res. Lett. **6**, 117 (1979)

Barton, M.J., Von Engel, A.: Phys. Rev. A **32**, 173 (1970)

Bartschat, K., Blum, K., Hanne, G.F., Kessler, J.: J. Phys. B **14**, 3761 (1981)

Bartschat, K., Blum, K.: Z. Phys. **304**, 85–88 (1982)

Bartschat, K., Fang, Y.: In: Madison, D.H., Schulz, M. Correlation, Polarization and Ionization in Atomic Systems, AIP Conf. Proc., vol. 604, pp. 202–209. American Institute of Physics, New York (2002)

Bartschat, K., Madison, D.H.: J. Phys. B **21**, 153–170 (1988)

Batelaan, H., van Heck, J., Heideman, H.G.M.: J. Phys. B **23**, 3993–4004 (1990)

Baum, G., Moede, M., Raith, W., Schroder, W.: In: Eichler, J., et al. (eds.) Proc. 13th Conf. on the Phys. of Elec. and Atomic Collisions, Berlin, pp. 144 (1983)

Becker, U., Shirley, D.A.: VUV and Soft X-Ray Photo-ionization, in Physics of Atoms and Molecules. Plenum, New York (1996)

Beckman, O.: Ark. Fys. **9**, 495 (1955)

Beijers, J.P.M., Madison, D.H., van Eck, J., Heideman, H.G.M.: J. Phys. B **20**, 167–181 (1987)

Beijers, J.P.M., van den Brink, J.P., van Eck, J., Heideman, H.G.M.: J. Phys. B **19**, L581–L585 (1986)

Bell, K.L., Kingston, A.E.: J. Phys. **88**, 2666 (1975)

Berezhko, E.G., Kabachnik, N.M., Sizov, V.V.: J. Phys. B **11**, 1819 (1978)

Bethe, H.: Ann. Phys. **5**, 325 (1930)

Bethe, H.: In: Flugge, S. (ed.) Handbuch der Physik, vol. 24, p. 1. Springer, Berlin (1933)

Beyer, H.F., Hippler, R., Shartner, K.-H., Albat, R.: Z. Phys. A **289**, 239 (1979), A292, 353 (1979)

Beyer, H.F., Kluge, H.J., Shevelko, V.P.: X-Ray Radiation of Highly Charged Ions. Springer, Berlin (1997)

Birks, L.S.: In: Herglotz, H.K., Birks, L.S. (eds.) X-ray Spectroscopy. Dakkar, New York (1978)

Blum, K., Kleinpoppen, H.: J. Phys. B **8**, 922 (1975)

Blum, K., Kleinpoppen, H.: Phys. Rep. **52**, 203 (1979)

Blum, K.: Density Matrix Theory and Application. Plenum, New York (1981)

Bonsen, T.F.M., Vriens, L.: Physica **47**, 307 (1970)

Born, M., Wolf, E.: Principles of Optics. Pergamon, New York (1970)

Brandt,W., Basbas, G: Phys. Rev. A **27**, 578 (1983)

Bransden, B.H., McDowell, M.R.C.: J. Phys. B **2**, 1187 (1969)

Bransden, B.H., McDowell, M.R.C.: J. Phys. B **3**, 29 (1970)

Bransden, B.H., McDowell, M.R.C.: Phys. Rep. **30c**, 209 (1977)

Bransden, B.H., McDowell, M.R.C.: Phys. Rep. **46**, 249 (1978)

Bray, I., Bartschat, K, Fursa, D.V, Stelbovics, A.T.: J. Phys. B **36**, 3425–3433 (2003)

Bray, I., Fursa, D.V., Stelbovics, A.T.: Phys. Rev. A **63**, 040702 (2001)

Bray, I., Fursa, D.V., Stelbovics, A.T.: Phys. Rev. A **74**, 034702 (2006)

Bray, I.: Phys. Rev. Lett. **89**(27), 273201–273204 (2002)

Bray, I., Fursa, D.V.: Phys. Rev. A **54**, 2991–3004 (1996)

Breinig, M., et al.: Phys. Rev. A **22**, 520–528 (1980)

Bridgeman, H.A.: J. Atmos. Chem. **12**, 299 (1991)

Brunger, M.J., Riley, J.L., Scholten, R.E., Teubner, P.J.O.: In: Proc. 15th Int. Conf. on the Physics of Electronic and Atomic Collisions. Brighton, Amsterdam, Abstracts pp. 170 (1987)

Brunger, M.J., Riley, J.L., Scholten, R.E., Teubner, P.J.O.: J. Phys. B **22**, 1431–1442 (1989)

Burgt, P.J.M., van der Antaya, M.E., McConkey, J.W.: Z. Phys. D **24**, 125 (1992)

Burhop, E.H.S.: The Auger Effect and Other Radiation-Less Transitions. Cambridge University Press, Cambridge (1952)

Burke, P.G., Schey, H.: Phys. Rev. A **126**, 163 (1962)

Cadez, I.M., Pejecv, V.M., Kurepa, M.V.: J. Phys. D **16**, 305 (1983)

Cairns, R.B., Harrison, H., Shoen, R.I.: Phys. Rev. **183**, 52 (1969)

Camilloni, R., Giadini, A., Tiribelli, R., Stefani, G.: Phys. Rev. Lett. **29**, 618 (1972)

Campbell, D.M., Brash, H.M., Farago, P.S.: Phys. Lett. **6**, 449 (1971)

Campbell, J.L.: Nucl. Instrum. Methods **65**, 333 (1968)

Carlson, T.A., Krause, M.O.: Phys. Rev. **137A**, 1655 (1965)

Carlson, T.A., Krause, M.O.: Phys. Rev. **151A**, 41 (1966)

Cartwright, D.C., Csanak, G.: J. Phys. B **19**, L485 (1986)

Chaudhry, M.A., Duncan, A.J., Hippler, R., Kleinpoppen, H.: Abstracts of Tenth Int. Conf. on Atomic Phys., p. 391, Tokyo, Japan, August 1986

Chaudhry, M.A., Duncan, A.J., Hippler, R., Kleinpoppen, H.: Phys. Rev. Lett. **59**, 2036 (1987)

Chaudhry, M.A., Kleinpoppen, H.: In: Kleinpoppen, H., Newell, W.R. (eds.) Polarized Electron/ Polarized Photon Physics, Proceedings of two United Kingdom Engineering and Physical Science Research Committee Workshops in Polarized Electron/Polarized Photon Physics held September, 1993 and April, 1994 in York, England. Plenum, New York

Chen, M.H., Craseman, B., Aoyagi, M., Huang, K.-N., Mark, H.: Atom. Data Nucl. Data Tables **26**, 561 (1981)

Chen, S.T., Gallagher, A.: Phys. Rev. A **14**, 593 (1976)

Chen, S.T., Leep, D., Gallagher, A.: Phys. Rev. A **13**, 947 (1976)

Cheng, A.F.: Astron. J. **242**, 812 (1980)

Cheng, W.O., Rudd, M.E., Hsu, Y.Y.: Phys. Rev. A **40**, 3599 (1989)

Cherid, M., Lahmam-Bennani, A., Duguet, A., Zarales, R.W., Lucchese, R.R., Dal Cappello, M.C., Dal Cappelo, C.J.: J. Phys. B **22**, 3483 (1989)

Cho, H., Hsieh, K.C., McIntyre, L.C. Jr.: Phys. Rev. A **33**, 2290 (1986)
Christensen R. L. et al.: Rev. Sci. Instr. **30**, 356 (1959)
Clark, R.E.H., Abdallah, J. Jr., Csanak, G., Kramer, S.P.: Phys. Rev. A **40**, 1935 (1989)
Clark, R.E.H., Abdallah, J. Jr.: Phys. Rev. A **44**, 2874 (1991)
Clark, R.E.H., Abdallah, J. Jr.: Private communication (1992)
Clausing, P.: Physica **9**, 65 (1929)
Cleff, B., Mehlhorn, W.: J. Phys. B **7**, 593 (1974)
Coburn, J.W.: Plasma Chem. Plasma Proc. **2**, 1 (1982)
Coglan, W.A., Clansing, R.E.: Atom. Data **5**, 317 (1973)
Colgan, J., Foster, M., Pindzola, M.S., Bray, I., Stelbovics, A.T., Fursa, D.V.: J. Phys. B **42**, 145002 (2009)
Colomb, P., Watanabe, K., Marmo, F.F.: J. Chem. Phys. **36**, 958 (1962)
Cooks, R.G., Terwilliger, D.T., Beynon, J.H.: J. Chem. Phys. **61**, 1208 (1974)
Cooper, G., Zarate, E.B., Jones, R.K., Brion, C.E.: Chem. Phys. **150**, 237 (1991a)
Cooper, G., Zarate, E.B., Jones, R.K., Brion, C.E.: Chem. Phys. **150**, 251 (1991b)
Cooper, J.W., Kolbenstvedt, H.: Phys. Rev. A **5**, 677 (1972)
Corney, A.: Atomic and Laser Spectroscopy. Clarendon, Oxford, p. 473 (1977)
Crowe, A., McConkey, J.W.: J. Phys. B **6**, 2088 (1973b)
Crowe, A., McConkey, J.W.: Phys. Rev. Lett. **31**, 192 (1973a)
Crowe, A., Rudge, M.R.H.: Comments on Atomic and Molecular Physics **22**(3), 147 (1988)
Csanak, G., Cartwright, D.C.: Phys. Rev. A **34**, 93–96 (1986)
Csanak, G., Cartwright, D.C.: Phys. Rev. A **36**, 2740 (1988)
Culhane, J.L., Fabian, A.C.: IEEE Trans. Nucl. Sci. **NS-19**, 569 (1972)
Culhane, J.L., Herring, J., Sandford, P.W., O'Shea, G., Phillips, R.D.: J.Sci. Instrum. **43**, 908 (1966)
Curtis, D.M., Eland, J.H.D.: Int. J. Mass Spectrom. Ion Proc. **63**, 241 (1985)
Dalgarno, A., McDowell, M.C.R.: In: Armstrong, E.B., Dalgarno, A. (eds.) The Airglow and the Aurorae. Pergamon, New York, p. 340 (1956)
De Heer, F.J., Jansen, R.H.J., Van der Kay, W.: J. Phys. B **12**, 979 (1979)
De Heer, F.J., Jansen, R.H.J.: J. Phys. B **10**, 3741 (1977)
Delwiche, J.: Bull. Acad. R. Belg. **55**, 215 (1969)
Dibeler, V.H., Mohler, F.L.: J. Res. Natl. Bur. Stand. **40**, 25 (1948)
Dibeler, V.H., Walker, J.A.: J. Chem. Phys. **44**, 4405 (1966)
Ding, H.B., Pang, W.N., Liu, Y.B., Shang, R.C., Chin. Phys. Lett. **22**(9), 2252 (2005)
Donnelly, B.P., Crowe, A.: Z. Phys. D **14**, 333–337 (1989)
Donnelly, B.P., Neill, P.A., Crowe, A.: J. Phys. B **21**, 1321–1325 (1988)
Dorman, F.H., Morrison, J.D.: J. Chem Phys. **35**, 575 (1961)
Drukarev, G.F.: Collision of Electrons with Atoms and Molecules. Plenum, New York, p. 76 (1987)
Dujardin, G., Besnard, M.J., Hellner, L., Malinovitch, Y.: Phys. Rev. A **35**, 5012 (1987)
Dunn, G.H., Kieffer, L.: J. Phys. Rev. **132**, 2109 (1963)
Dupre, C., Lahmam-Bennani, A., Duguet, A., Mota-Furtado, F., O'Mahony, P.F., Capello, C.D.: J. Phys. B **25**, 259–276 (1992)
Edmonds, A.R.: Angular Momentum in Quantum Mechanics. Princeton University Press, New Jersey (1959)
Edwards, A.K., Wood, R.M., Beard, A.S., Ezell, R.L.: Phys. Rev. A **37**, 3697 (1988)
Edwards, A.K., Wood, R.M., Davis, J.L., Ezell, R.L.: Nucl. Instrum. Methods Phys. Res. B **40/41**, 174 (1989)
Eggarter, E.: J. Chem. Phys. **62**, 833 (1975)
Ehlers, V.J., Gallagher, A.: Phys. Rev. A **7** 1573 (1973)
Ehrhardt, H., Hasselbacher, K.H., Jung, K., Schultz, M., Tekaat, T., Willman, K.: Z. Phys. **244**, 254 (1971)

Ehrhardt, H., Hasselbacher, K.H., Jung, K., Willman, K.: In: McDaniel, E.W., McDowell, M.R. C. (eds.) Case Studies in Atomic Collision Physics II, pp. 161–210. North-Holland, Amsterdam (1972a)

Ehrhardt, H., Schultz, M., Tekaat, T., Willman, K.: Phys. Rev. Lett. **22**, 89 (1969)

Elanbaas, W., Z. Phys. **59**, 289–305 (1930)

Eland, J.H.D., Wort, F.S., Lablanquie, P., Nenner, I.: (1986), Z. Phys. D **4**, 31

Eland, J.H.D., Wort, F.S., Royds, R.N.: J. Electr. Spectr. Rel. Phen. **41**, 297 (1986)

Eminyan, M., McAdam, K.B., Slevin, J., Kleinpoppen, H.: J. Phys. B **7**, 1519–1542 (1974)

Eminyan, M., McAdam, K.B., Slevin, J., Kleinpoppen, H.: Phys. Rev. Lett. 31, 576–579 (1973)

Eminyan, M., McAdam, K.B., Slevin, J., Standage, M.C., Kleinpoppen, H.: J. Phys. B **8**, 2058 (1975)

Endo, N., Kurogi, Y.: IEEE Trans. Electron Devices **27**, 1346 (1980)

Enemark, E.A., Gallagher, A.C.: Phys. Rev. A **6**, 192 (1972)

Erdman, P.W., Zipf, E.C.: Rev. Sci. Instrum. **53**, 225–227 (1982)

Fang, Y., Bartschat, K.: J. Phys. B **34**, L19–L25 (2000)

Fano, U., Macek, J.: Mod. Phys. **45**, 553 (1973)

Fano, U., Pan, X.C.: Comm. Atom. Mol. Phys. **26**(4), 203 (1991)

Fano, U.: Ann. Rev. Nucl. Sci. **13**, 1 (1963)

Fehsenfeld, F.C.: J. Chem. Phys. **53**, 2000 (1970)

Fink, R.W.: In: Craseman, B. (ed.) Atomic Inner-shell Processes II. Academic, New York (1975)

Flamm, D.L., Donnelly, V.M.: Plasma Chem. Plasma Proc. **1**, 317 (1981)

Fon, W.C., Berrington, K.A., Kingston, A.E., J. Phys. B **23**, 4347–4354 (1990)

Fon, W.C., Berrington, K.A., Kingston, A.E.: J. Phys. B **24**, 2160–2182 (1991)

Fon, W.C., Lim, K.P., Berrington, K.A., Lee, T.J.: J. Phys. B **28**, 1569–1583 (1995)

Fon, W.C., Lim, K.P., Sewey, P.M.J.: J. Phys. B **26**, L747–L752 (1993)

Forand, J.L., Becker, K., McConkey, J.W.: J. Phys. B **18** 1409–1418 (1985)

Franklin, J.L., Heirl, P.M., Whan, A.: J. Chem. Phys. **47**, 3148 (1967)

Fransiniski, L.J., Stankiewicz, M., Randall, K.J., Hatherly, P.A., Codling, K.: J. Phys. B At. Mol. Opt. Phys. **19**, L819 (1986)

Frees, L.C., Sauers, I., Ellis, H.W., Christophorous, L.G.: J. Phys. D Appl. Phys. **14**, 1629 (1981)

Friedburg, H.: Z. Phys. **130**, 493 (1951)

Fursa, D.V., Bray, I.: J. Phys. B **30**, 757–785 (1997)

Fursa, D.V., Bray, I.: Phys. Rev. A **52**, 1279–1298 (1995)

Gillespie, E.S.: J. Phys. B **5**, 1916–1921 (1972)

Giordmaine, J.A., Wang, T.C.: J. Appl. Phys. **31**, 463 (1960)

Glassgold, A.E., Lalongo, G.: Phys. Rev. **175**, 151 (1968)

Glassgold, A.E., Lalongo, G.: Phys. Rev. **186**, 266 (1969)

Gold, R., Bennet, E.F.: Phys. Rev. **147**, 201 (1966)

Goto, T., Hane, K., Okuda, M., Hattori, S.: Phys. Rev. A **28**, 1844–1850 (1983)

Gral, D., Fink, W.: J. Phys. B **18**, L803 (1985)

Green M. B. and Proca G. A. Rev. Sci. Instr. V **41**, 10, 1409–1414 (1970)

Hafid, H., Joulakian, B., Dal Cappello, C.: J. Phys. B **26**, 3415 (1993)

Haidt, D., Kleinpoppen, H.: Z. Phys. **196**, 72–76 (1966)

Haken, H., Wolf, H.C.: Atomic and Quantum Physics. Springer, Berlin (1984)

Halas, S.T., Adamczyk, B.: Int. J. Mass Spectrom. Ion Phys. **10**, 157 (1972/73)

Hamdy, H., Beyer, H.-J., Mahmoud, K.R., Zohny, E.I.M., Hassan, G., Kleinpoppen, H.: In: Proc. of 13th Int. Conf. on Atomic and Molecular Physics, Munich, Abstracts E23 (1992)

Hanle, W., Z. Phys. **30**, 93 (1942)

Hanssen, J., Joulakian, B., dall Cappello, C., Hafid, H.: J. Phys. B **27**, 3547 (1994)

Happer, W.: Atomic Physics, vol. 4. Plenum, New York, p. 651 (1975)

Harbach, T.: Dipl. Thesis, Fultat fur Physik, Universitat Bielefeld, Germany (1980)

Harland, P., Thynne, C.T.: J. Phys. Chem. **73**, 4031 (1969)

Hatch, E.N.: Z. Physik. **177**, 337 (1964)

Hauschild, W., Exner, W.: In: Gaseous Dielectrics V. Proc. Fifth Int. Symposium, Knoxville, TN, p. 419 (1987)

Hayes, P.A., Williams, J.F.: Phys. Rev. Lett. **77**, 3098 (1996)

Heddle, D., Gallagher, J.W.: Rev. Mod. Phys. **61**, 221 (1989)

Hille, E., Mark, T.Y.D.: J. Chem. Phys. **69**, 4600 (1978)

Hils, D., Jitschin, W., Kleinpoppen, H.: Appl. Phys. **25**, 39 (1981)

Hils, D.: Ph.D. Thesis, University of Colorado (1971)

Hippler, R., Bossler, J., Lutz, H.O.: J. Phys. B **17**, 245 (1984a)

Hippler, R., Jitchin, W.: Z. Phys. A **307**, 287 (1982)

Hippler, R., Saeed, K., Duncun, A.J., Kleinpoppen, H.: Phys. Rev. A **30**, 5 (1984b)

Hippler, R., Saeed, K., Mc Gregor, I., Kleinpoppen, H.: Phys. Rev. Lett. **56**, 1622 (1981)

Hippler, R.: In: Beyer, H.J., Kleinpoppen, H. (eds.) Progress in Atomic Spectroscopy (part C). Plenum, New York (1984)

Hitchcock, A.P., Brion, C.E., Van der Wiel, M.J.: J. Phys. B At. Mol. Phys. **11**, 3245 (1978)

Hitchcock, A.P., Van der Wiel, M.J.: J. Phys. B At. Mol. Phys. **12**, 2153 (1979)

Hughes, R.W., Weaver, L.D.: Phys. Rev. **132**, 710 (1963)

Hughes, V.W., Long, R.L., Lubell, M.S., Posner, M., Raith, W.: Phys. Rev. A **5**, 195 (1972)

Humphery, I., Williams, J.F., Heck, E.L.: J. Phys. B **20**, 367–391 (1987)

Ibrahiem, K.S., Beyer, H.J., Kleinpoppen, H.: In: Proc. 14th Int. Conf. on Physics of Electronic and Atomic Collisions (Palo Alto), p. 118. North Holland, Amsterdam, Abstracts (1985)

Ibrahiem, K.S.: Ph.D. Thesis, University of Stirling, Scotland (1986)

Igual-Ruiz, N., Donnelly, B.P., McLaughlin, D.T., Cvejanovic, D., Crowe, A., Fursa, D., Bartschat, K., Bray, I.: J. Phys. B **34**, 2289 (2001)

Inaba, S., Hane, K., Goto, T.: J. Phys. B **19**, 1371–1376 (1986)

Inokuti, M., Kim, Y.K., Platzman, R.L.: Phys. Rev. **164**, 55 (1967)

Inokuti, M.: Rev. Mod. Phys. **43**, 297 (1971)

Isozumi, Y., Isozumi, S.: Nucl. Instum. Methods **96**, 317 (1971)

Jacobowiccz, H., Moores, D.L.: In: Juergen, H. (ed.) Electron–Atom and Electron–Molecule Collisions. Plenum, New York (1983)

Jain, D.K., Khare, S.P.: J. Phys. B **8**, 1429 (1967)

Jenkin, R., de Vrie, J.L.: Practical X-ray Spectroscopy, London: Macmillan (1970)

Jitschin, W., Osimitsch, S., Reihe, H., Kleinpoppen, H., Lutz, H.: J. Phys. B At. Mol. Opt. Phys **17**, 1899 (1984)

Jitschin, W., Wisotzki, B., Werner, U., Lutz, H.O.: J. Phys. E Sci Instrum. **17**, 137 (1984)

Johnstone, W.M., Newell, W.R.: J. Phys. B At. Mol. Opt. Phys. **24**, 473 (1991)

Kennedy J.V., Muerscough, V.P., Mc Dowell, M.R.: J. Phys. B At. Mol. Opt. Phys. **10**, 3759 (1977)

Khakoo, M.A., Becker, K., Forand, J.L., McConkey, J.W.: J. Phys. B **19**, L209–L213 (1986)

Khiefets, A.S., Bray, I., Bartchat, K.: J. Phys. B **33**, L433–L438 (1999)

Kieffer, L.J., Dunn, G.H.: Phys. Rev. **158**, 61 (1967)

Kieffer, L.J., Dunn, G.H.: Rev. Mod. Phys. **38**, 1 (1966)

Kim, Y.K., Inokuti, M.: Phys. Rev. A **7**, 1257 (1973)

Kim, Y.K., Stephan, K., Mark, E., Mark, T.D.: J. Chem. Phys. **74**, 6771 (1981)

Kim, Y.K.: Phys. Rev. A **28**, 656 (1983)

Kleinpoppen, H., Lehmann, B., Grum-Grzhimailo., Becker, U.: In: Advances in Atomic, Molecular, and Optical Physics, vol. **51**, p. 471 (2005)

Kleinpoppen, H.: Phys. Rev. A **3**, 2015 (1971)

Kleinpoppen, H.: Physics of One or Two Electron Atoms, p. 612. North-Holland, Amsterdam (1969)

Kluge, H.J., Sauter, H.: Z. Phys. **270**, 295–309 (1974)

Knudson, A.R. et al.: Phys. Rev. Lett. **37**, 679 (1976)

Kohmoto, M., Fano, U.: J. Phys. B **14**, L447–L451 (1981)

Kokta, L.: Nucl. Instrum. Methods **112**, 245 (1973)

Kopferman, H.: Kernmomente. Academic, New York (1958)

Kossman, H., Schwarzkopf, O., Schmidt, V.: J. Phys. B **23**, 301 (1990)

Kuhlenkemp, H.: Z. Phys. **157**, 282 (1959)

Kuyatt, C.E.: Rev. Sci. Instr. **39**, 33 (1968)

Krause, M.O.: In: Wuilleumier, F. (ed.) Photoionization and Other Probes of Many-Electron Interactions, p. 133. Plenum, New York (1976)

Krause, M.O.: J. Chem. Phys. Ref. Data **8**, 307 (1979)

Kuhlenkempf, H., Schmidt, L.: Ann. Phys. **43**, 494 (1943)

Lahmam-Bennani, A., Dupre, C., Duguet, A.: Phys. Rev. Lett. **63**, 1582 (1989)

Lahmam-Bennani, A., Wellinstein, H.F., Duguet, A., Rouault, M.: J. Phys. B **16**, 121 (1983)

Lahmam-Bennani, A.: J. Phys. B **24**, 2401 (1991)

Lange, M., Matsumoto, J., Lower, J., Buckman, S., Zatsarinny, O., Bartschat, K., Bray, I., Fursa, D.: J. Phys. B **39**, 4179 (2006)

Laudau, L.D., Lifshitz, E.M.: Quantum Mechanics Non-Relativistic Theory, p. 579. Pergamon, Oxford (1965)

Lee, E.T.P., Lin, C.C.: Phys. Rev. **138**, A301–A304 (1965)

Leep, D., Gallagher, A.: Phys. Rev. A **13**, 148–155 (1976)

Lewyn, L.L.: Nucl. Instrum. Methods **82**, 138 (1970)

Lifshitz, C., Tiernan, T.O., Hughes, B.M.: J. Chem. Phys. **59**, 3183 (1973)

Loch, S.D., Pindzola, M.S., Balance, C.P., Griffin, D.C., Mitnik, D.M., Badnell, N.R., O'Mullane, M.G., Summers, H.P., Whiteford, A.D.: Phys. Rev. A **66**, 052708 (2002)

Lucas, C.B.: J. Phys. E **6**, 991 (1973)

Lucchese, R.R., Dal Cappello, M.C., Dal Capppello, C.J., J. Phys. B **22**, 3483 (1989)

Macek, J., Jaeck, J.D.: Phys. Rev. A **4**, 2288 (1971)

Madison, D.H., Csanak, G., Cartwright, D.C.: J. Phys. B **19**, 3361–3366 (1986)

Madison, D.H.: J. Phys. B **12**, 3399 (1979)

Madison, D.H.: Phys. Rev. **8**, 2449–2455 (1973)

Madison, D.H.: Private communication (1982)

Madison, D.H., Shelton, W.N.: Phys. Rev. A **7**, 499 (1973)

Manson, S.T., Tuburen, L.H., Madison, D.H., Stolterfoht, N., Phys. Rev. A **12**, 60 (1975)

Margreiter, D., Deutsch, H., Schmidt, M., MärkT.D.: Int. J. Mass Spectrom. Ion Proc. **100**, 157 (1990b)

Märk, T.D., Dunn, G.H.: Electron impact ionization. In: Märk, T.D., Dunn, G.H. (eds.) Ch. 5, Ionization of SF6 by Electron Impact. Springer, New York (1985)

Mark, T.D.: J. Chem. Phys. **63**, 3731 (1975)

Marriott, J.: Thesis, University of Liverpool, England (1954)

Martus, K.E., Becker, K., Madison, D.H., Phys. Rev. A **38**, 4876 (1988)

Massey, H.S.W., Burhop, E.H.S., Gilbody, H.B.: Electronic and Ionic Impact Phenomenon, 2nd ed. Clarendon, Oxford (1969)

Massey, H.S.W., Mohr, C.B.O.: Proc. R. Soc. A **136**, 289–311 (1932)

Massey, H.S.W., Mohr, C.B.O.: Proc. R. Soc. A **140**, 613 (1933)

Masuoka, T., Samson, J.A.R.: J. Chem. Phys. **75**, 4946 (1981)

Masuoka, T.: J. Chem. Phys. **98**, 6989 (1993)

Mathis, R.F., Vroom, D.A.: J. Chem. Phys. **64**, 1146 (1976)

McColloh, K.E.: J. Chem. Phys. **48**, 2090 (1968)

McGuire, E.J.: Phys. Rev. A **11**, 1880 (1975)

McGuire, E.J.: Phys. Rev. A **9**, 1840 (1974)

McGuire, J.H.: Phys. Rev. Lett. **49**, 1153 (1982)

Mehlhorn, W.: In: Wuilleumier, F. (ed.) Photo-Ionization and Other Probes of Many-Electron Interactions, p. 309. Plenum, New York (1976)

Mehlhorn, W.: Z. Phys. **208**, 1 (1968)

Melisse, L., Moody, S.E.: Rev. Sci. Instrum. **48**, 131 (1977)

Meneses, G.D., Pagan, C.B., Machado, L.E.: Phys. Rev. A **41**, 4610 (1990)

Mitroy, J., McCarthy, I.E.: J. Phys. B **22**, 641–654 (1989)

Moisewitch, B.L., Smith, S.J.: Rev. Mod. Phys. **140**, 238 (1969)
Mokler, P.H., Folkmann, F.: In: Sellin, I.A. (ed.) Structure and Collisions of Ions and Atoms, p. 201. Springer, Berlin (1978)
Moores, D., Golden, L.B., Sampson, P.: J. Phys. B **13**, 385 (1980)
Moores, D.L., Norcross, D.W.: J. Phys. B **5**, 1482 (1972)
Morita, S.: Research Report, Institute of Plasma Physics, Nagoya University, Nagoya, Japan (1983)
Moruzzi, M.: In: Moruzzi, G., Strumia, F. (eds.) The Hanle Effect and Level Crossing Spectroscopy, pp. 34–46. Plenum, New York (1991)
Mott, N.F., Massey, B.S.W.: The Theory of Atomic Collisions, 3rd ed., p. 489. Clarendon, Oxford (1965)
Nagy, P., Skutlartz, A., Schmidt, V.: J. Phys. B **13**, 1249 (1980)
Nakel, W.: Radiat. Phys. Chem. **75**(10), 1164 (2006)
Nier, A.O.: Int. J. Mass Spectrom. Ion Phys. **66**, 55 (1985)
Oda, N., Nishimura, F., Tahira, S.: J. Phys. Soc. Japan **33**, 462 (1972)
Ogurtsov, G.N.: Sov. Phys. JETP **37**, 584 (1973)
Oldham, W.J.B.: Phys. Rev. A **140**, 1477 (1965)
Oldham, W.J.B.: Phys. Rev. A **161**, 1 (1967)
Omidvar, K., Kyle, H.L., Sullivan, E.C.: Phys. Rev. **45**, 1124 (1972)
Oona, H.: Phys. Rev. Lett. **32**, 571 (1974)
Opal, C.B., Beaty, E.C., Peterson, W.K.: Atom. Data **4**, 209 (1972)
Opal, C.B., Peterson, W.K., Beaty, E.C.: J. Chem. Phys. **55**, 4100 (1971)
Orient, O.J., Srivastava, S.K.: J. Chem. Phys. **80**, 140 (1984)
Orient, O.J., Srivastava, S.K.: J. Phys. B **20**, 3923 (1987)
Osimitsch, S.: Diplomarbeit., Universitat Bielefeld (1983)
Papp, F.F., Omanjuk, N., Shpenik, C.D.: In: ICPEAC XIII: Abstracts of contributed papers, p. 744, Berlin, Germany (1983)
Parrat, L.G.: Phys. Rev. **50**(1), 13 (1936)
Penetrante, B.M., Bardsely, J.N., Vitello, P.A., Vogtlin, G.E., Hofer, W.W.: 44th Gaseous Electronics Conference, Abstracts, p. 159, Albuquerque, New Mexico (1991)
Percival, I.C., Seaton, M.: J. Phil. Trans. R. Soc. A **251**, 113–138 (1958)
Peresse, M.M.J., Tuffin, F.: Methods Physiques D'Analyse (GAMS), Jan–Mar (1967)
Perlman, H.S.: Proc. R. Soc. (Lon.) **76**, 623 (1960)
Peterkop, R.: Theory of Ionization of Atoms by Electron Impact, Colorado Associated University Press, Colorado (1977)
Peterson, W.K., Opal, C.B., Beaty, E.C.: J. Phys. B **4**, 1020 (1971)
Peterson, W.K., Opal, C.B., Beaty, E.C.: Phys. Rev. A **5**, 712 (1972)
Phelps, J.O., Lin, C.: Phys. Rev. A **24**, 1299 (1981)
Philip, L.B., Stelbovics, A.T.: Phys. Rev. A **66**, 012707 (2002)
Pinto, R., Ramanthan, K.V., Babau, R.S.: J. Electrochem. Soc. **134**, 165 (1987)
Pratt, R.H., Tseng, H.K., Lee, C.M., Kissel, L., Mc Callum, C., Riley, M.: Atom. Data Nucl. Data Tables **20**, 175 (1977)
Proca, G.A., Green, M.B. Rev Sci. Instr. V **41**, (12), 1778–1783 (1970)
Pullen, B.P., Stockdale, J.A.D.: J. Int. Mass Spectrom. Ion Phys. **21**, 35 (1976)
Ramsey, N.F. in "Molecular Beams" Oxford University Press, England (1956)
Ramsey, N.F. et al.: Phys. Rev. Lett. **23**, 1369 (1969)
Rapp, D., Englander-Golden, P.: J. Chem. Phys. **43**, 1464 (1965)
Ravon, J.P.: Nucl. Istrum. Methods **211**, 7 (1982)
Reese, R.M., Dibeler, V.H., Franklin, J.L.: J. Chem. Phys. **29**, 880 (1958)
Robb, W.D.: J. Phys. B **7**, 1006 (1974)
Roque, M.B., Siegel, R.B., Martus, K.E., Tanovsky, V., Becker, K.: J. Chem. Phys. **94**, 341 (1991)
Röntgen, W.C.: Nature **53**, 274 (1896)
Rubin, K., Perl, J., Bederson, B.: Phys. Rev. **117**, 151 (1960)

Rudd, M.E., Hollman, K.W., Lewis, J.K., Johnson, D.L., Porter., R., Fagerquist, E.L.: Phys. Rev. A **47**, 1866 (1993)

Rudd, M.E., Sautter, C.A., Bailey, C.L.: Phys. Rev. A **151**, 20 (1966)

Rudd, M.E.: Phys. Rev. A **44**, 1644 (1991)

Rudge, M.R.H.: Rev. Mod. Phys. **40**, 564 (1968)

Sadlej, A.J., Urban, M., Gropen, O.: Phys. Rev. A **41**, 5547 (1991)

Saxon, R.P.: Phys. Rev. A **8**, 839 (1973)

Schmieder, R.W., Lurio, A., Happer, W., Khadjavi, A.: Phys. Rev. A **2**, 1216, A173, 76 (1970)

Schmitz, W., Mehlhorn, W.: J. Phys. E **5**, 64 (1972)

Schnopper, H.W.: Phys. Rev. **154**, 118 (1966)

Schram, B.L., De Heer, F.J., Van der Wiel, M.J., Kistemaker, J.: Physica, **31**, 94 (1965)

Schram, B.L.: Physica **32**, 197 (1966)

Schröder, W.: Ph.D. Thesis, University of Bielefeld, Germany (1982)

Schultz, D.R., Meng, L., Olson, R.E.: J. Phys. B At. Mol. Opt. Phys. **25**, 4601 (1992)

Sewell, E.C., Crowe, A.: J. Phys. B **15**, L357 (1982)

Shah, M.B., Gilbody, M.B.: J. Phys. B **15**, 3453 (1982)

Shergin, A.P., Gordeev, Y.u.S.: In: Invited Papers and Progress Reports of X Int. Conf. on 'The Physics of Electronic and Atomic Collisions', p. 377, Paris (1977)

Shintarou, K., Takeshi, K., Chauhan, R.J., Srivastava, R., Shinobu, N.: J. Phys. B **39**, 493–503 (2006)

Short, R.T.C.-S.O., Levin, J.C., Sellin, I.A., Liljiby, L., Huldt, S., Johansson, S.E., Nilsson, E., Church, D.A.: Phys. Rev. Lett. 56(2614) (1986)

Shyn, T.W., Sharp, W.E., Kim, Y.K.: Phys. Rev. A **24**, 79 (1981)

Shyn, T.W., Sharp, W.E.: Phys. Rev. A **19**, 557 (1979a)

Shyn, T.W., Sharp, W.E.: Phys. Rev. A **20**, 2332 (1979b)

Shyn, T.W., Sharp, W.E.: Phys. Rev. A **43**, 2300 (1991)

Simpson, J.A., Kuyatt, C.E.: J. Res. Natl. Bur. Stand. **67C**, 279 (1963)

Simpson, J.A., Kuyatt, C.E.: Rev. Sci. Instrum. **34**, 265 (1963)

Simpson, J.A.: Methods of Experimental Phys. 4(part A), p. 84 (1967)

Smith, O.I., Stevenson, J.S.: J. Chem. Phys. **74**, 6777 (1981)

Smyth, H.D., Muller, P.W.: Phys. Rev. **43**, 121 (1933)

Sneddon, I.N., Fourer Transforms, p. 365. McGraw Hill, London (1951)

Srivastava, R., Zuo, T., McEachran, R.P., Stauffer, A.D.: J. Phys. B **25**, 3709–3720 (1992)

Soa, E.A.: Jenaer Jahrbuch. **1**, 115 (1959)

Standage, M.C., Kleinpoppen, H.: Phys. Rev. Lett. **36**, 577–580 (1976)

Stanski, T., Adamczyk, B.: Int. J. Mass Spectrom. Ion Phys. **46**, 31 (1983)

Stelbovics, A.T.: Phys. Rev. Lett. **83**(8), 1570–1573 (1999)

Steph, N.C., Golden, D.E.: Phys. Rev. A **21**, 759–770 (1980)

Stevenson, M., Crowe, A.: J. Phys. B **37**, 2493–2500 (2004)

Stolterfoht, N., de Heer, F.I., Van Eck, J.: Phys. Rev. Lett. **30**, 1159 (1973)

Tahira, S., Oda, N.: J. Phys. Soc. Japan 35, 582. In: Proc. of the 8th Int. Conf. on 'The Physics of Electronic and Atomic Collisions', Belgrade, pp. 443–463 (1973)

Talib, Z.A., Saporoschenko, M.: Int. J. Mass Spectrom. Ion Proc. **116**, 1 (1992)

Tawara, H., Itikawa, Y., Nishimura, H., Yoshino, M.: J. Phys. Chem. Ref. Data. **19**, 617 (1990)

Taylor, P.O., Dunn, G.H.: Phys. Rev. A **8**, 2304–2321 (1973)

Tripathi, A.N., Mathur, K.J.: J. Phys. B **6**, 1431 (1973)

Tseng, H.K., Pratt, R.H., Lee, C.M.: Phys. Rev. A **13**, 187 (1979)

Urey, G.T.: Phys. Rev. **11**, 401 (1918)

Van Brunt, R.J., Kieffer, L.: Phys. Rev. A **2**, 1293 (1970)

Van der Weil, N.J., Wiebs, G.: Physica **53**, 225 (1971)

Vriens, L.: Physica **45**, 400 (1969)

Vriens, L.: Physica **47**, 267 (1970)

Vroom, D.A., Palmer, R.L., McGowan, J.W.M.: J. Chem. Phys. **66**, 647 (1977)

Vucie, S., Potvlige, R.M., Joachain, C.J.: Phys. Rev. A **35**, 1446 (1987)

Weissbluth, M.: Atoms and Molecules, p. 340. Academic, New York (1978)

Wendin, G., Ohno, M.: Phys Scr **14**, 148 (1976)

Wentzel, G.: Z. Phys. **43**, 524 (1927)

Werij, H.G.C., Greene, C.H., Theodosiou, C.E., Gallagher, A.: Phys. Rev. A **46**, 1248–1260 (1992)

Werner, U.: Dipl. Thesis, Facultat fur Physik, Universitat Bielefeld, Germany (1983)

Wetzel, W.W.: Phys. Rev. **44**, 25 (1933)

Wille, U., Hippler, R.: Phys. Rep. **132**, 3–4 (1986)

Williams, J.F., Humphrey, I.: In: Coggiola, M.J., Huestis, D.H., Saxon, R.P. (eds.) Proc. 14th Int. Conf: on the Physics of Electronic and Atomic Collisions, Palo Alto, p 112. North Holland, Amsterdam (1985)

Williams, J.F.: In: Eichler, J., et al. (eds.) Proc. 13th Int. Conf: on the Physics of Electronic and Atomic Collisions (Bedin), Abstracts, p. 132. North-Holland, Amsterdam (1983)

Worthington, C.R., Tomlin, S.G.: Proc. Phys. Soc. A **69**, 401 (1956)

Zetner, P.W., Trajmar, S., Csanak, G.: Phys. Rev. A **41**, 5980 (1990)

Ziezel, J.P.: J. Chem. Phys. **64**, 595 (1967)

Zipf, E.C.: In: Mark, T.D., Dunn, G.H. (eds.) Electron Impact Ionization. Springer, Vienna (1985)

Zohny, E.I.M., El-Fayoumi, M.A.K., Hamdy, H., Beyer, H.J., Eid, Y. Shahin, F., Kleinpoppen, H.: In: Proc. 16th Int. Conf. on Physics of Electronic and Atomic Collisions, New York, p. 173. North Holland, Amsterdam (1989)

# Index